MATHEMATICAL MODELING WITH EXCEL®

The Jones and Bartlett Publishers Series in Mathematics

The Way of Analysis, Revised Edition
Strichartz (978-0-7637-1497-0) © 2000

Topology

Foundations of Topology, Second Edition
Patty (978-0-7637-4234-8) © 2009

Discrete Math and Logic

*Discrete Structures, Logic, and Computability,
Third Edition*
Hein (978-0-7637-7206-2) © 2010

Essentials of Discrete Mathematics
Hunter (978-0-7637-4892-0) © 2009

Logic, Sets, and Recursion, Second Edition
Causey (978-0-7637-3784-9) © 2006

Numerical Methods

Numerical Mathematics
Grasselli/Pelinovsky (978-0-7637-3767-2) © 2008

*Exploring Numerical Methods: An Introduction to
Scientific Computing Using MATLAB*
Linz (978-0-7637-1499-4) © 2003

Advanced Mathematics

Mathematical Modeling with Excel
Albright (978-0-7637-6566-8) © 2010

*Clinical Statistics: Introducing Clinical Trials,
Survival Analysis, and Longitudinal Data Analysis*
Korosteleva (978-0-7637-5850-9) © 2009

Harmonic Analysis: A Gentle Introduction
DeVito (978-0-7637-3893-8) © 2007

Beginning Number Theory, Second Edition
Robbins (978-0-7637-3768-9) © 2006

A Gateway to Higher Mathematics
Goodfriend (978-0-7637-2733-8) © 2006

*For more information on this series and its titles, please visit us online at
http://www.jbpub.com/math. Qualified instructors, contact your Publisher's Representative at
1-800-832-0034 or info@jbpub.com to request review copies for course consideration.*

The Jones and Bartlett Publishers International Series in Mathematics

MATHEMATICAL MODELING WITH EXCEL®

BRIAN ALBRIGHT, DSc

Assistant Professor of Mathematics
Department of Mathematics and Computer Science
Concordia University, Nebraska

JONES AND BARTLETT PUBLISHERS

Sudbury, Massachusetts

BOSTON   TORONTO   LONDON   SINGAPORE

World Headquarters

Jones and Bartlett Publishers
40 Tall Pine Drive
Sudbury, MA 01776
978-443-5000
info@jbpub.com
www.jbpub.com

Jones and Bartlett Publishers
Canada
6339 Ormindale Way
Mississauga, Ontario L5V 1J2
Canada

Jones and Bartlett Publishers
International
Barb House, Barb Mews
London W6 7PA
United Kingdom

Jones and Bartlett's books and products are available through most bookstores and online booksellers. To contact Jones and Bartlett Publishers directly, call 800-832-0034, fax 978-443-8000, or visit our website, www.jbpub.com.

Substantial discounts on bulk quantities of Jones and Bartlett's publications are available to corporations, professional associations, and other qualified organizations. For details and specific discount information, contact the special sales department at Jones and Bartlett via the above contact information or send an email to specialsales@jbpub.com.

Production Credits

Publisher: David Pallai
Acquisitions Editor: Timothy Anderson
Editorial Assistant: Melissa Potter
Production Director: Amy Rose
Production Manager: Jennifer Bagdigian
Production Assistant: Ashlee Hazeltine
Senior Marketing Manager: Andrea DeFronzo
V.P., Manufacturing and Inventory Control:
    Therese Connell

Composition: Vikatan Publishing
Art Rendering: Vikatan Publishing
Figure 1.1 Rendering: Matthew Moore
Title Page and Cover Design: Kristin E. Parker
Cover Image: © Comstock Images/age fotostock
Printing and Binding: Malloy, Inc.
Cover Printing: Malloy, Inc.

Library of Congress Cataloging-in-Publication Data
Albright, Brian, 1977-
    Mathematical modeling with Excel / Brian Albright.
        p. cm.
        Includes bibliographical references and index.
        ISBN-13: 978-0-7637-6566-8 (hardcover)
        ISBN-10: 0-7637-6566-X (hardcover)
    1. Mathematical analysis–Data processing. 2. Mathematical models–Data processing.
    3. Microsoft Excel (Computer file) I. Title.
        QA300.A446 2010
        511'.8–dc22

                                                                            2009010208

6048
Printed in the United States of America
13 12 11 10 09   10 9 8 7 6 5 4 3 2 1

Preface

The main goal of *Mathematical Modeling with Excel®* is to present different ways of building and analyzing mathematical models in a format that can be read by students, not only instructors. This is not a text on how to use Excel. Rather, Excel is seen as a tool to further the goal of building and analyzing mathematical models. No prior knowledge or experience with Excel is required to use this text.

Excel is chosen as the only software used to implement and analyze models for two main reasons:

1. It is easy to use and most everyone is familiar with it, so it takes very little time to become comfortable with the software.
2. It is everywhere. Students will have access to Excel for every mathematical modeling project they encounter inside and outside of academics.

Each section contains step-by-step instructions for building the models in Excel. These instructions were originally written for use with Microsoft Office Excel 2003. Some of the instructions may be slightly different for other versions of Excel.

Pedagogical Approach

This text presents a wide variety of common types of models found in other mathematical modeling texts, as well as some new types. However, the models are presented in a very unique format. A typical section begins with a general description of the scenario being modeled after which the model is built using the appropriate mathematical tools. Then, it is implemented and analyzed in Excel via step-by-step instructions. In the exercises, we ask students to modify or refine the existing model, analyze it further, or adapt it to similar scenarios.

In each section, we try to focus on the main mathematical modeling concept being illustrated and not get too bogged down in details. We also focus on the analysis of models, and in each case try to address the question "What does this mean?"

This is not a "plug-and-chug" textbook. We do not ask students to simply plug numbers into some "black-box" Excel formula and accept the results. Rather, we discuss the mathematics behind the analysis of the models and, where appropriate, build the analytical tools in Excel from scratch.

Each section ends with several exercises of varying degrees of difficulty. In addition, each chapter ends with a "For Further Reading" section that contains resources for additional information.

Audience/Prerequisites

This text is appropriate for mathematics majors (including secondary mathematics education majors) who need an introductory mathematical modeling course. Some sections require calculus, linear algebra, differential equations, or basic statistics, so this text is appropriate for use with junior- or senior-level students. However, many other sections require only mathematical maturity, so this text could also be used with sophomore-level students.

The Flow of Material

This text contains a wide variety of modeling techniques, mathematical concepts, and types of applications. There is more than a semester's worth of material in this text. An instructor can easily pick and choose the sections that are appropriate for the students.

Chapter 1—What Is Mathematical Modeling? This chapter begins with the definitions of the terms *model* and *mathematical modeling*. It then discusses the steps involved in mathematical modeling, and concludes with a discussion of the importance of assumptions in the process of mathematical modeling.

Chapter 2—Proportionality and Geometric Similarity. This chapter opens with an introduction to graphing and working with data in Excel, including a discussion of fitting straight lines to data. Then the geometric concepts of proportionality and similarity are used to model systems such as free-falling objects. We stress the point that data is used to test the validity of the models.

Chapter 3—Empirical Modeling. This chapter begins with a discussion of fitting models, such as exponential and power, to data by transforming the data. Then the coefficient of determination is introduced as a way to measure how well a model fits a set of data. Ways of using linear algebra and least-squares solutions to fit polynomials and other types of models to data are discussed. The chapter ends with a discussion of spline models.

Chapter 4—Discrete Dynamical Systems. We open this chapter with the definitions of a discrete dynamical system, a solution, and an equilibrium value. We stress the point that we are usually interested in the long-term behavior of the system, not necessarily the value at a single point in time, and how equilibrium values are important in the analysis. The chapter includes several types of applications modeled with discrete dynamical systems, including population growth, predator–prey systems, and simple epidemics.

Chapter 5—Differential Equations. This chapter begins with a discussion of the fact that it is often easier to describe how a quantity changes over time than it is to describe the value of the quantity at any particular time. This motivates the use of differential equations for modeling dynamical systems. We focus on finding approximate numerical solutions to differential equations, rather than finding exact analytical solutions. To this end, we discuss Euler's method for approximating solutions to differential equations and apply it to systems of differential equations. The chapter concludes with a discussion of eigenvalues and how they are related to the behavior of a system.

Chapter 6—Simulation Modeling. We cover the topic of simulations more extensively than most other mathematical modeling text books. The main goal of this chapter is to illustrate several different types of simulation models, including games of chance, queuing models, inventory models, and scheduling models. We also discuss how pseudo-random number generators work, as well as how to model random variables using density functions.

Chapter 7—Optimization. The main focus of this chapter is linear programming and the simplex method. We do not discuss much theory; rather we try to give students an overview of the basic ideas behind the simplex method. We introduce the assignment problem and the transportation problem as examples of linear programs and how to model with those programs. We also briefly introduce the gradient method to approximate solutions to nonlinear programs.

Additional Information

For answers, worksheets, and the author-created Excel worksheet, "Linear Programming," please visit http://www.jbpub.com/catalog/9780763765668.

Acknowledgments

The author would like to thank the following reviewers, who provided valuable suggestions during the development of this book:

- Bruce DeGroot, Concordia University, Seward, NE
- Gary DeYoung, Dordt College, Sioux Center, IA
- Greg Grindey, United States Transportation Command, St. Louis, MO
- Len Cabrera, United States Air Force Academy, Colorado Springs, CO
- Philip Gustafson, Mesa State University, Grand Junction, CO

Contents

CHAPTER 1

What Is Mathematical Modeling?

Chapter Objectives

- Define the terms *model*, *mathematical model*, and *mathematical modeling*
- Understand the purpose and process of mathematical modeling
- Understand the importance and significance of assumptions behind mathematical models

Every student of mathematics has done some mathematical modeling in his or her educational career. These instances of mathematical modeling are typically called *applications* and are used to illustrate how mathematics is implemented in the real world.

In most math classes, the main goal is to learn the theory of some particular mathematical discipline. The applications are used to help achieve this goal by providing a more concrete context in which to study and understand the theory. For instance, in Calculus I, the goal is to understand the idea of the limit and the derivative. An applied maximization problem is used to motivate the idea of the derivative and to provide practice in calculating and interpreting derivatives.

In mathematical modeling, the opposite is true. Here we start with some real-world problem and use mathematical theory and techniques to better understand the phenomena behind the problem.

1.1   Definitions

To define the phrase *mathematical modeling*, we first define the term *model*. The word *model* is used frequently in everyday language. We talk about model airplanes, model houses, models on a runway, and so on. What does the term *model* mean in a mathematical sense?

Lucas (1999, 5) defines a model as "a simpler realization or idealization of some more complex reality." The real world is a very complex place. To better understand it, we need to try to simplify it to a reasonable degree, describe the simplification in ways we can understand and work with, and then study the simplification. This is what we call *modeling*.

A *mathematical model* then can be defined as a model constructed using mathematical terms, symbols, and ideas. Giordano (2003, 54) defines a mathematical model as "a mathematical construct designed to study a particular real-world system or phenomenon." Mathematical models can take many different forms. They may involve equations, inequalities, differential equations, matrices, logic, or any other type of "mathematical" idea.

The key idea is that we use mathematics to describe a portion of the real world. Therefore, a very simple but general definition of the *process* of mathematical modeling is

Definition 1.1.1 *Mathematical modeling* is the application of mathematics to real-world problems.

1.2   Purpose

Why do we do mathematical modeling? If we want to answer a question about real-world phenomena, we could just sit back, observe, and take notes. Suppose we put 500 bacteria in a Petri dish. The next day we count 525 in the dish, and the next we count 551.

Obviously, the number of bacteria is growing. Based on this observation, we might ask these questions:

1. How long will it be until there are 600 bacteria in the dish?
2. If we need 900 bacteria for an experiment in 3 days, how many must we put into the dish today?

We could answer each question as follows:

1. Wait until we count 600 bacteria in the dish.
2. Put 1 bacterium in a dish, 2 in a second dish, 3 in a third, and so on, up to 900; wait 3 days, and determine which dish contains 900 bacteria.

These solutions only require us to make simple observations of this real-world phenomenon of bacteria growing in a Petri dish. However, these solutions are obviously impractical, because they might require too much time or too many resources (the second solution requires 900 Petri dishes and a total of $1 + 2 + \cdots + 900 = 405{,}450$ bacteria).

A much more practical approach to answering these questions is to construct a function that gives the number of bacteria in the dish in terms of time (i.e., construct a mathematical model of the bacteria growth).

In other situations, making observations may itself be a complicated ordeal. For instance, suppose we want to find the optimal mixture of doctors and nurses (i.e., the number of doctors and the number of nurses) to staff a hospital emergency room. The concept of "optimal" may take into account several factors, including

1. Quality of patient care. (Do patients get the care they need?)
2. Patient waiting time. (Do patients have to wait a long time?)
3. Time spent with patients. (Are the doctors and nurses overworked, or do they have too much "free time"?)
4. Resources. (Is there enough floor space or are people running into each other?)

One approach to finding an optimal number is to choose some mixture of doctors and nurses (say three doctors and eight nurses), put them to work, and have a team of people record data for a series of weeks or months. Then choose another mixture (say two doctors and seven nurses) and repeat the process. Repeat this until all possible combinations of doctors and nurses have been tried, analyze the data, and pick the optimal mixture.

This approach has many of the same problems as the bacteria growth problem. It would take too much time and be too expensive. Plus there are additional problems. If there are too few doctors and nurses on staff, for example, patients might unnecessarily die. Also, we may not observe how the different mixtures handle infrequent events such as a bus crash that floods the emergency room with dozens of patients at once.

A more practical solution would be to try to replicate the behavior in the emergency room on a computer (i.e., create a type of mathematical model called a simulation) where the numbers of doctors and nurses can be easily changed. Each mixture can be simulated for a long period of time under many different situations and at low cost. Plus, nobody dies.

1.3   The Process

We illustrate the process of mathematical modeling with an example of modeling the number of bacteria in a Petri dish as described in Section 1.2.

   Step 1: State the question to be answered

In many situations, this step is almost trivial; in others, it is the most difficult part of the process. The question should be narrow enough to make the problem manageable, but not too narrow so that the problem is trivial. Initially we may want to focus on a narrow question, and then use the knowledge gained to broaden the question at a later time. The question should also be stated in precise mathematical terms so it can easily be translated into mathematical notation.

In this example we answer the question, "How long will it be until the number of bacteria in the dish reaches 600?"

   Step 2: Select the modeling approach

In this step we determine the form of the model. In some situations this is easy to do; in others we may have several reasonable choices. Making the right choice requires at least some knowledge of all the possibilities. It also depends on the nature of the assumptions being made.

Oftentimes this step begins with some simple observations. Note that we started with 500 bacteria. After 1 day, the number increased by 25, which is 5% of 500. After a second day, it increased by 26, which is approximately 5% of 525. The growth rate (or change per day) appears to be relatively constant. This suggests a simple relationship between the populations on consecutive days:

$$\text{Population on one day} = \text{Population on the previous day} + 5\%$$

This relationship indicates that we may be able to derive a simple equation to model the population.

Step 3: Define variables and parameters

Variables are quantities that could change within a problem. *Parameters* are quantities that are constant within a problem, but could change between problems of the same type. The first part of this step is to determine what variables and parameters are involved. This may be simple and obvious, or very complicated. Oftentimes there are potentially hundreds of quantities involved. To make the model manageable, we need to make assumptions as to which are the most important and which can be ignored. At a later time we can add additional variables and parameters to refine the model.

In this example, variables include

1. Time
2. Population
3. Temperature
4. Amount of food present
5. Amount of available space in the dish

These are all values that change as the population grows. Since the initial observation did not give any information on temperature, food, or space, we ignore these variables and focus only on time and population.

Possible parameters to consider include

1. The initial population
2. Growth rate (we assume this is constant)
3. Size of the dish
4. Initial amount of food

These are all values that are constant once we put the bacteria in the dish and allow them to grow. But if we consider a different dish with a different population of bacteria, these values could change. Again, since we don't know anything about the size of the dish or the amount of food, we ignore these parameters.

The second part of this step is to choose symbols to represent the variables and parameters. For this example, let

$$n = \text{time in days from the present } (n = 0, 1, ...)$$
$$r = \text{the growth rate (in decimal form)}$$
$$a_n = \text{the population at the beginning of day } n$$
$$a_0 = \text{the initial population.}$$

Step 4: State the assumptions

Making assumptions is an essential aspect of creating a valid and manageable model. Assumptions fall into many different categories. Some are used to simplify the model, such as those used to select the important variables. Some are needed to define relationships between the variables because the precise relationships are not known. Others are needed to determine the values of parameters when the exact values are not known.

Clearly stating the assumptions is an important part of interpreting and presenting the results. The results of a model are only as valid as the underlying assumptions. If the assumptions are unreasonable, then the conclusion will be unreasonable regardless of the precision of the mathematical analysis.

In this problem, we have already chosen to ignore temperature, size of the dish, and many other possible variables and parameters. This is a simplification. Furthermore, we assume that the population growth is constant (i.e., the population will increase 5% each day).

Step 5: Formulate the model

This is where the "mathematics" starts. We have observed that the number of bacteria on day 1 is equal to the number on day 0 plus 5%. The number on day 2 is equal to the number on day 1 plus 5%, and so on. In mathematical notation using our variables and parameters, we have

$$a_1 = a_0 + r\, a_0 = (1 + r)\, a_0$$
$$a_2 = a_1 + r\, a_1 = (1 + r)\, a_1$$
$$\vdots$$
$$a_{n+1} = a_n + r\, a_n = (1 + r)\, a_n.$$

This forms a recursively defined sequence. To form an explicit description of a_n in terms of n, note that

$$a_1 = (1 + r)\, a_0$$
$$a_2 = (1 + r)\, a_1 = (1 + r)(1 + r)\, a_0 = (1 + r)^2 a_0$$
$$a_3 = (1 + r)\, a_2 = (1 + r)(1 + r)^2\, a_0 = (1 + r)^3 a_0$$
$$\vdots$$
$$a_n = (1 + r)^n a_0.$$

This last equation is our model.

Step 6: Solve the model and state the solution

Here we loosely use the term *solve*. Solving a model may involve solving a single equation, as in this example. In other models it may involve constructing a graph and qualitatively describing its behavior, or it may involve running a simulation several hundred times and summarizing the resulting data. The meaning of the term *solve* is relative to the type of model.

In this example, the question is "When will the population be 600?" In terms of our variables, this can be stated as "Find n such that $a_n = 600$." This yields the equation

$$600 = (1 + 0.05)^n 500.$$

Solving this equation using logarithms yields $n \approx 3.7$. This means that at the beginning of the fourth day we will have over 600 bacteria. This is our solution.

Oftentimes the results of a model are used to guide decisions. In many practical situations, such as in business or the military, the person doing the modeling is not the final decision maker. The final decision maker is a chief executive officer or officer who is not a mathematician. Therefore, the solution should be stated in as nontechnical language as reasonably possible.

Step 7: Verify the model

Verification is necessary to test the reasonableness of our assumptions. Typically we verify a model by comparing it to some real-world data. Let's suppose we let the bacteria grow for a total of 7 days and collect the data in Table 1.1. Next to the actual observed populations are the populations predicted by the model.

Day	Actual Population	Predicted Population
0	500	500
1	525	525
2	551	551
3	575	579
4	598	608
5	610	638
6	620	670

Table 1.1

We see that on day 4, the actual population is just below 600. Even though the predicted population does not equal the actual population, our solution of 4 days is reasonable.

Note that on days 5 and 6, the actual and predicted populations differ considerably. The data indicates that the growth rate slows down. This means that our assumption of a constant growth rate is incorrect.

Our model is accurate up to day 4, but inaccurate for later days. This example illustrates that we must be very cautious about using data from the past to make predictions about the future.

Step 8: Refine the model

Refine means to improve the model in some way. One way to do this is to add variables that we chose to ignore in Step 3 to make a more accurate model. Another way is to generalize the model so that it can be used to solve other similar problems. Either approach will require us to repeat Steps 3–7 to some degree.

We have already noted that the data indicates the growth rate slows down over time. This could be a result of diminishing food supplies or room to grow. These are two variables we chose to ignore.

One possible refinement is to redo the model incorporating these two variables. This would require additional observations and data to determine how these variables are related to the other variables. Another possible refinement is to use the available data to model a decreasing growth rate. We illustrate how to do this in Chapter 2.

A simple diagram that illustrates the basic process of mathematical modeling is given in Figure 1.1. The process begins in the upper-left corner with observations (or data). From this we get the basic problem we want to solve. We then make assumptions, construct the model, solve it using appropriate mathematical tools, and obtain a mathematical result. Then we must interpret the mathematical result in light of the assumptions to make our conclusions. We then verify the model using more observations.

This figure also illustrates the cyclic nature of mathematical modeling. We rarely stop once we answer the original question. We continually repeat the process, to some extent, to test, refine, and implement the model.

The right half of this diagram is done in the "math world" and the left half is done in the "real world." In the math world, we use the absolute certainty of mathematics. The real world contains no such certainties. Making assumptions and interpreting are necessary steps to move between these two worlds.

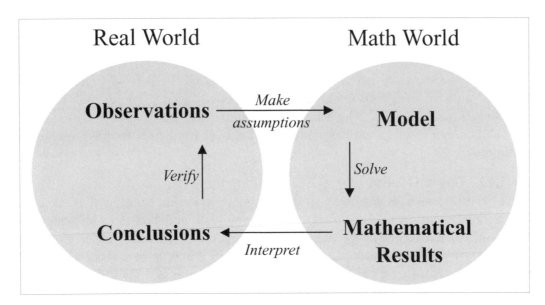

Figure 1.1

1.4 Assumptions

Every model is based on some set of assumptions. Sometimes those assumptions are rather trivial and obvious, but most times they are significant enough to potentially affect the validity of the model. We will never be able to describe each component of a real-world system exactly. Assumptions are needed to fill in these gaps, and whenever possible, the reasonableness of assumptions should be tested. In fact, assumptions are so important that any mention of a model should include the assumptions behind it. A definition of mathematical model that takes this into account follows.

Definition 1.4.1 A *mathematical model* is a mathematical interpretation of assumptions concerning real-world problems.

Here are a few examples of well-known models and some of their underlying assumptions:

Example 1.4.1 EPA Fuel Mileage
When the Environmental Protection Agency (EPA) tests the fuel economy of new vehicles, model it (or simulates) the driving habits and conditions of a typical driver. Since every driver is different, these cannot be modeled exactly, so they are scripted (i.e., assumed). Prior to 2008, the scripts were based on data from the 1970s! (In 1984, the calculations were

slightly modified to lower the estimates. See the EPA website at http://www.epa.gov/fuel economy/420f06009.htm for more details.) As a result, according to *Consumer Reports* (Consumers Union of the United States, 2006), these estimates were inaccurate in 90% of new cars and overestimated actual fuel mileages by as much as 50%. These inaccurate numbers not only affected consumers, but also affected national energy policies that are formed using these estimates. The scripts were changed in 2008 to better model current driving habits. As a result, identical models carried over from 2007 to 2008 received lower mpg estimates.

Example 1.4.2   Carbon-14 Dating
Carbon-14 dating methods are used to date organic material, such as a piece of bone, found at archeological sites. The underlying mathematical model requires knowledge of the proportion of carbon-14 originally present in the sample. Obviously we cannot measure this precisely, so we assume that this proportion is the same as in a modern bone. This assumption can't be tested directly, but if a date can be confirmed via independent means, it would be an indication that the assumption is correct.

Example 1.4.3   Newtonian Mechanics
Newton's second law says that the force exerted on an object is equal to its rest mass times the acceleration, or $F = ma$. This was assumed to hold at any velocity. In the 20th century, it was discovered that for speeds approaching the speed of light, this rest mass must be replaced with the relativistic mass, which is larger.

Example 1.4.4   Relativity
Einstein's theory of special relativity actually is a mathematical model. It is based on several postulates (a type of assumption), one of which is that the speed of light in a vacuum is constant to any observer in an inertial frame of reference. These assumptions cannot be tested in every circumstance imaginable, but the results of Einstein's theory can be tested. These results have shown to be correct, which suggests that the underlying assumptions are also correct.

Exercises

Directions: Identify at least one type of assumption behind the model used in each situation:

1.4.1 A consumer magazine reports that a laundry detergent costs $0.31 per load.

1.4.2 An exercise bike displays the number of calories burned.

1.4.3 A jogging pedometer measures the distance jogged.

1.4.4 A dashboard display in a car shows that it can travel 220 mi on the remaining fuel in the tank.

1.4.5 A small business owner predicts that his company will spend $500,000 on phone bills next year.

1.4.6 The manufacturer of a battery for a radio-controlled airplane claims that you will get 12 min of flight time on a single charge.

For Further Reading

- For a further discussion of the modeling process and related issues, see F. R. Giordano, M. D. Weir, and W. P. Fox. 2003. *A first course in mathematical modeling.* 3rd ed. Pacific Groves, CA: Thomson Brooks/Cole, 52–63.
- For more information on modeling methodology, tips on writing reports, and numerous example problems, see D. Edwards and M. Hamson. 1990. *Guide to mathematical modelling.* Boca Raton, FL: CRC Press.
- For a discussion of the benefits of mathematical modeling, see W. F. Lucas. 1999. The impact and benefits of mathematical modeling. In *Applied mathematical modeling*, ed. D. R. Shier and K. T. Wallenius, 1–25. Boca Raton, FL: Chapman and Hall/CRC.
- For a lengthy discussion of how assumptions impact environmental models, see O. H. Pilkey and L. Pilkey-Jarvis. 2007. *Useless arithmetic: Why environmental scientists can't predict the future.* New York, NY: Columbia University Press.
- For a brief discussion of some of the assumptions behind global-warming models, see K. K. Tung. 2007. *Topics in mathematical modeling.* Princeton, NJ: Princeton University Press, 146–157.

References

Consumers Union of the United States. 2006. Fuel economy: Why you're not getting the mpg you expect. *Consumer Reports*, June.

Giordano, F. R., M. D. Weir, and W. P. Fox. 2003. *A first course in mathematical modeling.* 3rd ed. Pacific Groves, CA: Thomson Brooks/Cole.

Lucas, W. F. 1999. The impact and benefits of mathematical modeling. In *Applied mathematical modeling*, ed. D. R. Shier and K. T. Wallenius, 5. Boca Raton, FL: Chapman and Hall/CRC.

CHAPTER 2

Proportionality and Geometric Similarity

Chapter Objectives

- Use *proportionality* and *geometric similarity* as simplifying assumptions in the modeling process
- Fit straight lines to data
- Use data to find constants of proportionality
- Introduce the basics of graphing with Excel

2.1   Introduction

One of the first steps in modeling is to make simplifying assumptions, and one of the most basic types of assumptions is that one variable is simply a constant multiple of the other. This type of relationship is called *proportionality*.

Definition 2.1.1 The variable y is said to be *proportional* to the variable x if there exists a nonzero constant c (called the *constant of proportionality*) such that

$$y = cx. \tag{2.1}$$

The expression $y \alpha x$ is used to indicate that y is proportional to x. Oftentimes the phrase "directly proportional" is used to describe this type of relationship. We must point out that a proportionality relationship between two variables does not mean that one variable *causes* the other.

Note that if $y = cx$, then $x = \frac{1}{c}y$ so that x is proportional to y. Thus, if y is proportional to x, we immediately know that x is proportional to y (i.e., a proportionality relationship is symmetric). For this reason, if y is proportional to x, we simply say that x and y are proportional.

Graphically, $y \alpha x$ means that graph of y versus x (y on the vertical axis and x on the horizontal axis) should form a straight line through the origin. The constant of proportionality is the slope of this line.

2.2   Using Data

One famous proportionality relationship is *Hooke's law*, which relates the force applied to a spring to the distance it is stretched or compressed. Hooke's law simply states that

$$d = kF \tag{2.2}$$

where F is the force applied to a spring, d is the distance stretched or compressed, and k is a constant related to the stiffness of the spring.

Example 2.2.1   Bucket on a Spring

Suppose that we hang a bucket from a spring, fill the bucket with varying amounts of sand, and measure the distance the spring is stretched. The results are recorded in Table 2.1.

We plot this data to (1) verify that Hooke's law holds for this spring and (2) find the constant of proportionality. This data comes from the real world, so it is subject to errors and uncertainty. Therefore, we cannot expect the data to lie perfectly on a straight line as

Weight (newtons)	Distance (cm)
5	1.02
10	1.86
15	3.00
20	3.94
25	4.95
30	5.82
35	6.95
40	7.80

Table 2.1

predicted by the idealized Hooke's law. However, the data should lie very near a straight line. The slope of this line is the constant of proportionality.

In this and following examples we provide instructions for producing worksheets that allow us to explore the mathematical models we will be building throughout the rest of this text. As you become more familiar with building worksheets you will be expected to be able to construct worksheets for models with fewer instructions.

1. Rename a blank worksheet "**Spring**." Format the worksheet as in Figure 2.1 and enter the rest of the data from Table 2.1.

2. Follow these steps to form a scatter plot as in Figure 2.2:
   (a) Highlight the data (including the headings).
   (b) Click on the **Chart Wizard** button in the toolbar. (The **Chart Wizard** button looks like this: 📊 . If you do not see this button, select **Insert** → **Chart....**)
   (c) Select **XY (scatter)** under **Chart type**.
   (d) Choose the top **Chart sub-type** and click **Next** twice.
   (e) On the **Titles** tab, enter the **value x-axis** as **Weight** and **value y-axis** as **Distance**.
   (f) Click **Next** and then **Finish**.

	A	B
1	**Weight**	**Distance**
2	5	1.02
3	10	1.86
4	15	3.00

Figure 2.1

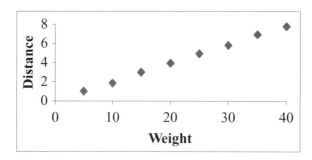

Figure 2.2

(g) Left-click on the legend and press **Delete**.
(h) Left-click on the plot area and press **Delete**.
(i) Left-click on the horizontal gridlines and press **Delete**.
(j) Right-click on the x-axis and select **Format Axis** ... .
(k) Under **Scale**, set the **minimum** to **0** and **maximum** to **40** and click **OK**.
(l) Right-click on the y-axis and select **Format Axis** ... .
(m) Under **Scale**, set the **minimum** to **0** and **maximum** to **8** and click **OK**.

3. Next we need to estimate the slope of a line that "fits" these points. If we were doing this with paper and pencil we would use a ruler or straightedge to draw a line through the origin that goes close to each data point and then find the slope. We will do the electronic version of this in Excel. Format the spreadsheet as in Figure 2.3.

	D	E	F
1	**Line**		
2	**x**	**y**	**Slope**
3	0	0	0.1
4	40	=F3*D4	

Figure 2.3

4. Next we will add these points in Figure 2.3 to the graph.

(a) Highlight the range **D3:E4**, left-click and hold on the box surrounding this range, and drag it onto the graph.
(b) Under **Add cells as**, select **New Series**.
(c) Under **Values (Y) in**, select **Columns**.
(d) Check the box next to **Categories (X value) in First Column**.
(e) Right-click on one of these new points and select **Chart type**. Select the type **Scatter with data points connected with lines** in the lower-left corner and click **OK**. Your graph should look similar to Figure 2.4.

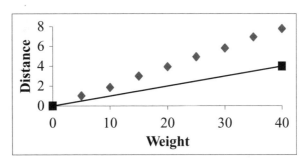

Figure 2.4

5. Change the value of the slope in cell **F3**. Notice that the graph of the line automatically changes.

6. Next we will "move" the line so it better fits the data and find the slope of this line. Follow these steps:

   (a) Left-click on the data point on the right end of the line, release, wait 2 seconds, and then click again. The cursor should change to a cross.
   (b) Left-click again and hold. You should now be able to move the data point up and down.
   (c) Move it so that the line appears to go through the data points reasonably well. Release the left-mouse button and a **Goal seek** window should appear.
   (d) Next to **By Changing Cell:**, select cell **F3** and click **OK**.
   (e) Your graph should look similar to Figure 2.5 and the value of the slope in cell **F3** should be approximately 0.195. (**Note:** Goal seek calculated the value of slope that gives the y-value of the point you moved the right end of the line to. Your value for the slope may be different.)

Conclusion: We see that the data does indeed lie very near a straight line, so Hooke's law is verified (at least in this example). The constant of proportionality is approximately 0.195. Therefore, if we know the amount of weight w on the spring we can approximate the distance the spring has been stretched, d, by

$$d = 0.195w.$$

Likewise, if we know the distance the spring has been stretched we can approximate the amount of weight on the spring by

$$w = \frac{1}{0.195}d \approx 5.128\,d.$$

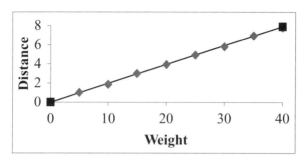

Figure 2.5

Example 2.2.2

A mathematics professor claims that the grade on the final exam is directly proportional to the amount of time spent studying. To test this claim, he asked each student how much time he or she studied and recorded it along with the grade. A graph of the collected data is shown in Figure 2.6.

Notice that the points do not lie close to a straight line through the origin, so the grade is not directly proportional to time spent studying. In fact, the points appear to be randomly scattered. This indicates that there is not much of a relationship at all between the two variables.

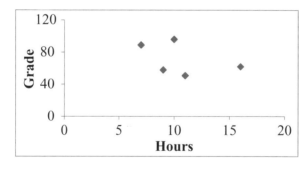

Figure 2.6

Example 2.2.3   Boyle's Law

Another well-known proportionality relationship is Boyle's law, which relates the volume of a gas to its pressure at a constant temperature,

$$V = \frac{k}{P}$$

where V denotes the volume of the gas, P denotes its pressure, and k is a constant. To test Boyle's law, a student measures the pressure of a gas at different volumes while keeping the temperature of the gas constant. The resulting data is shown in Table 2.2.

Volume (V)	50	45	40	35	30	25	20	15	10
Pressure (P)	27.24	30.36	34.01	38.73	45.5	54.35	68.03	90.53	135.73

Table 2.2

We will use this data to verify Boyle's law and find the constant of proportionality for this gas. Note that the law does not say that V is proportional to P. It says that V is proportional to $\frac{1}{P}$. Therefore we will plot V versus $\frac{1}{P}$ and fit a straight line through the origin to this *transformed* data.

1. Rename a blank worksheet "**Boyle**" and format it as in Figure 2.7. Enter the data from Table 2.2 in columns **A** and **C**. Left-click on cell **B2**. Double-click on the small box in the lower-right corner of the border. The formula in **B2** will be copied down to row 10.

	A	B	C
1	P	1/P	V
2	27.24	=1/A2	50

Figure 2.7

2. Use the transformed data in column **B** and the original data in column **C** to form a scatter plot as in Figure 2.8.

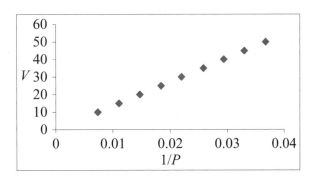

Figure 2.8

3. We see in Figure 2.8 that the data lie very near a straight line, as expected. To estimate the slope of this line, we simply pick one of the data points and calculate the slope of the line through the origin and the point. Let's choose the rightmost point $(0.036711, 50)$. The slope of a line through this point and the origin is

$$\text{slope} = \frac{50 - 0}{0.036711 - 0} \approx 1362.$$

4. To examine how well a line through the origin with this slope fits the data, format the spreadsheet as in Figure 2.9.

	E	F	G
1	Line		
2	x	y	Slope
3	0	0	1362
4	0.036711	=G3*E4	

Figure 2.9

5. Add the line to the scatter plot. Your graph should look similar to Figure 2.10.

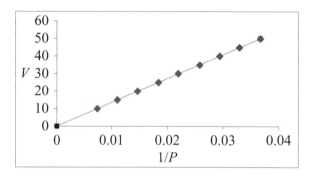

Figure 2.10

Conclusion: We see that V is clearly proportional to $\frac{1}{P}$ and the constant of proportionality is near 1362. Note that the constant of proportionality may be different for a different gas.

Exercises

2.2.1 For each of the data sets below, determine if it is reasonable to assume that y is proportional to x. If it is, approximate the constant of proportionality. If it is not, describe why this assumption is not reasonable.

1.

x	1	1.1	1.2	1.3	1.4	1.5	1.6	1.7
y	1	1.21	1.44	1.69	1.96	2.25	2.56	2.89

2.

x	1	5	7	2	10	12	3	6
y	0.79	10.89	14.37	5.75	23.36	26.29	3.76	16.12

3.

x	2	6	9	15	7	25	39	4
y	26	20	18	26	6	19	20	13

2.2.2 Suppose you drive your car on a perfectly flat road at a constant speed with no wind. In this case, the amount of fuel, y, (in gallons) needed is directly proportional to the distance traveled, x (in miles).

1. If the distance traveled increases, what can we say about the amount of fuel needed?

2. If the relationship is given by $y = 0.04x$ and x increases by 50 mi, how much does y increase?

3. Now, suppose it takes 12 gal of fuel to travel 282 mi. Find the constant of proportionality.

4. In words, describe the meaning of this constant of proportionality.

2.2.3 A variable y is said to be inversely proportional to x if there exists a constant c such that $y = \frac{c}{x}$.

1. If y is inversely proportional to x, sketch what a graph of y versus x would look like. What would a graph of y versus $1/x$ look like?

2. If y is inversely proportional to x, show that x is inversely proportional to y.

3. If y is inversely proportional to x and x increases, what happens to the value of y?

2.2.4 For the data set below, determine if it is reasonable to assume that y is inversely proportional to x. If it is, approximate the constant of proportionality. If it is not, describe why this assumption is not reasonable.

x	1	1.2	1.4	1.6	1.8	2.0	2.2	2.4
y	6.85	6.21	4.24	4.32	3.92	3.18	2.93	2.96

2.2.5 Newton's law of universal gravitation states that the force of attraction F between two objects with masses m_1 and m_2 is proportional to the product of the masses and inversely proportional to the square of the distance d between them. In mathematical notation,

$$F = k\frac{m_1 m_2}{d^2}.$$

For two given objects, if m_1 and m_2 are constant, we may combine them with the constant of proportionality k to describe the relationship by

$$F = \frac{C}{d^2}$$

where C is a constant. If one of the two objects is a planet, the distance d is the distance from the center of the planet to the second object.

1. The radius of the Earth is approximately 4000 mi. A satellite weighs 15 tons on the surface of the Earth (i.e., the force of attraction between the Earth and the satellite is 15 tons at the surface of the Earth). Use this information to calculate the constant of proportionality C.

2. Find the weight of the satellite (in tons) at an altitude of 500 mi above the surface of the Earth.

3. Use Excel to graph the weight of the satellite versus altitude for values of altitude between 0 and 4000 mi (ignore the effects of all other celestial bodies like the Moon).

2.2.6 Determine which of the following models "best" fits the data below. Find the constant of proportionality for this model. Explain why your choice is the "best" model.

$$y \alpha x, \quad y \alpha \frac{1}{x}, \quad y \alpha x^2, \quad y \alpha \sqrt{x}, \quad y \alpha \frac{1}{x^3}$$

x	0.5	0.7	0.9	1.2	1.5
y	7.8	3.5	2.2	0.85	0.36

2.2.7 Prove each of the following properties of proportionality:

1. If $ab \alpha ac$ and $a \neq 0$, then $b \alpha c$.

2. If $a^m \alpha ac$ and $a \neq 0$, then $a^{m-1} \alpha c$.

3. If $a \alpha c^m$, then $a^{1/m} \alpha c$.

2.2.8 A snow-cone seller at a county fair wants to model the number of cones he will sell, C, in terms of the daily attendance a, the temperature T, the price p, and the number of other food vendors n. He makes the following assumptions:

1. C is directly proportional to a and T is greater than 85°F.

2. C is inversely proportional to p and n.

Derive a model for C consistent with these assumptions. For what values of T is this model valid?

2.3   Modeling with Proportionality

One important observation about a proportionality relationship is that if one of the variables increases, so does the other, and if one variable decreases, so does the other. This can be seen in Figure 2.2 and Figure 2.8. Whenever we encounter a situation where two variables increase or decrease at the same time, we should consider a proportionality relationship.

Example 2.3.1   Population Growth

In many situations involving populations, the larger the population, the faster it grows. This suggests a proportionality relationship between the population and the rate of growth. Table 2.3 shows the population of bacteria in a Petri dish at different points in time. The third column contains the change in population between time periods.

Observe that as n increases, p_n increases, and so does Δp_n. This suggests that p_n is proportional to Δp_n. A graph of Δp_n versus p_n is shown in Figure 2.11. Here we see that

Day (n)	Actual Population (p_n)	Change in Population $(\Delta p_n = p_{n+1} - p_n)$
0	10.3	6.9
1	17.2	9.8
2	27.0	18.3
3	45.3	34.9
4	80.2	45.1
5	125.3	50.9
6	176.2	79.4
7	255.6	

Table 2.3

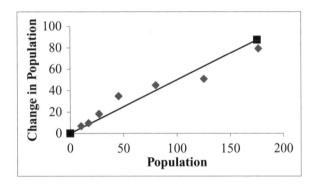

Figure 2.11

the points fall near a straight line through the origin, verifying that proportionality is a reasonable assumption. The slope of this line is approximately 0.5. This gives a model that relates the population at one day, p_n, to the population at the next, p_{n+1}:

$$\Delta p_n = p_{n+1} - p_n = 0.5p_n \quad \Rightarrow \quad p_{n+1} = 1.5p_n.$$

This model predicts that the population grows by about 50% during each time period, which means the population will grow without bound. This seems unreasonable, so the model needs to be refined. We will do this in Chapter 4.

Example 2.3.2   Radioactive Decay

One-half of the amount of a radioactive substance decays after each half-life. Radioactive carbon-14 (^{14}C) has a half-life of 5715 years. If we start with 10 g of ^{14}C, Table 2.4 shows the amount of material remaining after each half-life along with the rate of change between time periods.

Time (years)	Amount	Change
0	10.0	5.0
5715	5.0	2.5
11,430	2.5	1.25
17,145	1.25	0.625
22,860	0.625	

Table 2.4

Note that as the amount of ^{14}C decreases, the rate at which it decreases also changes. This suggests a proportionality relationship between the amount of ^{14}C and the rate at which it decreases.

If we let $y(t)$ represent the amount of ^{14}C at time t, this proportionality relationship gives the differential equation

$$\frac{dy}{dt} = ky.$$

Solving this differential equation yields the exponential model for growth $y(t) = Ce^{kt}$ where C is the initial amount of material.

In the previous examples we used data of some type to suggest a proportionality relationship. Oftentimes, as in the next example, we simply use logic to assume a proportionality relationship.

Example 2.3.3   Free-Falling Object

An object in free-fall encounters two basic forces. The first is its weight due to gravity. The second is air resistance that slows the rate of fall. Air resistance is typically negligible at low speeds so it is often not modeled. If we ignore air resistance, then the only force acting on the object is acceleration due to gravity. This leads to the simple differential equation

$$\frac{dv}{dt} = g$$

where $v(t)$ = velocity of the object at time t and $g = 9.8\,\mathrm{m/s^2}$ (the acceleration due to gravity). Solving this differential equation yields the model $v(t) = gt + v_0$ where v_0 is the initial velocity. This model predicts that the velocity grows without bound, which is inaccurate. Physics tells us that a free-falling object reaches a "terminal velocity" where the deceleration due to air resistance equals the acceleration due to gravity. At this point the object remains at a fairly constant velocity.

To refine this model for velocity, we can incorporate a simple model for air resistance. It seems reasonable to assume that as velocity increases, the force due to air resistance increases. This implies a proportionality relationship between velocity and the force due to air resistance:

$$\text{force due to air resistance} = kv.$$

This force acts upward on the object. There is also a force acting downward on the object due to its mass m:

$$\text{downward force} = mg.$$

Now, by Newton's second law,

$$F = ma = m\frac{dv}{dt}. \tag{2.3}$$

Also,

$$F = \text{downward force} - \text{upward force.} \tag{2.4}$$

Putting equations (2.3) and (2.4) together we get

$$m\frac{dv}{dt} = mg - kv$$
$$\Rightarrow \quad \frac{dv}{dt} + \frac{k}{m}v = g.$$

Solving this last differential equation gives the model

$$v(t) = \frac{mg}{k}\left(1 - e^{-kt/m}\right).$$

Note that in this model,

$$\lim_{t \to \infty} v(t) = \frac{mg}{k}.$$

This suggests a terminal velocity, so this model is more realistic.

Proportionality relationships satisfy the following transitive property:

Theorem 2.3.1 *If $a \, \alpha \, b$ and $b \, \alpha \, c$, then $a \, \alpha \, c$.*

Proof: By definition, $a \, \alpha \, b$ and $b \, \alpha \, c$ mean that $a = k_1 b$ and $b = k_2 c$ for some nonzero constants k_1 and k_2. So, substituting we get

$$a = k_1 k_2 c$$

but $k_1 k_2$ is a nonzero constant, so $a \, \alpha \, c$ by definition. $\square$

The next example illustrates an application of this property.

Example 2.3.4   Work Done by a Train

If a constant force is applied to an object in moving it some distance, the work done is defined to be

$$\text{work} = \text{force} \times \text{distance}.$$

Suppose a train engine pulls a car along a flat stretch of track until it runs out of fuel, and assume that the force needed is constant. We want to model the total work done by the engine in terms of the amount of fuel it carries.

Since the force is constant, work is proportional to the distance pulled. Let W denote work and D denote distance. In standard notation,

$$W \, \alpha \, D.$$

Now, the total distance the train can pull the car is related to the amount of fuel. The more fuel, the farther it can pull the car. So distance pulled is proportional to the amount of fuel. If A denotes the amount of fuel, we have

$$D \, \alpha \, A.$$

Combining these two proportionality relationships, we arrive at the model

$$W \, \alpha \, A.$$

Although we do not know the constant of proportionality, we can use this relationship to find relative values. For instance, if the constant of proportionality were 10 and the tank holds 1000 gal, with a full tank the train could perform

$$W = 10(1000) = 10,000$$

units of work. If the fuel tank were one-quarter full, it could perform

$$W = 10(250) = 2500$$

units of work, which is exactly one-quarter as much work as if the tank were full.

Exercises

2.3.1 A very simple assumption for the population of rabbits in a forest states that it grows at a rate proportional to the size of the population and the rabbits die at a rate proportional to the number of foxes in the forest. If p_n denotes the population of rabbits at time n, and F denotes the (constant) number of foxes, this assumption yields a model for the change in the rabbit population:

$$\Delta p_n = p_{n+1} - p_n = k_1 p_n - k_2 F$$

where k_1, $k_2 \geq 0$.

1. Solve this model for p_{n+1}.

2. If $p_0 = 500$, $k_1 = 0.15$, and $k_2 = 0.25$, find the values of F for which the population of rabbits is decreasing, for which it is increasing, and for which it is constant. (**Hint:** If the population is constant, $p_1 = p_0$.)

3. Create a spreadsheet to support your answer above.

2.3.2 Refer back to Example 2.3.1. We will analyze a refined model. Suppose we assume that Δp_n is proportional to the product of the population and its difference from 621, that is,

$$\Delta p_n = p_{n+1} - p_n = k p_n (621 - p_n).$$

1. Use the data given in Table 2.3 to test this assumption by plotting Δp_n versus $p_n(621 - p_n)$. Use the graph to estimate the constant of proportionality.

2. Use Excel to find the values of the population for hours 0 through 20 using this model (use the same initial population as in Table 2.3).

3. Does this model seem more or less reasonable than the original one? Why or why not?

2.3.3 A student wishes to determine how the size of a cloud affects the speed of a falling raindrop. She comes up with the following assumptions:

1. The speed is proportional to the weight of the raindrop.

2. The weight is proportional to the size, or volume, of the raindrop.

3. The size is proportional to the size of the cloud.

Use these assumptions to model the speed in terms of the size the cloud. Don't forget to define the variables. Does this model seem reasonable? Why or why not?

2.3.4 Prove the following statement: If a is inversely proportional to b and b is directly proportional to c, then a is inversely proportional to c.

2.3.5 Martin wants to determine how the amount of money he has in his wallet will affect his grade in Math Modeling class. Consider the following assumptions:

1. His grade is directly proportional to the amount of time studied.

2. The amount of time studied is directly proportional to the amount of free time he has.

3. The amount of free time he has is inversely proportional to the amount of time he spends going out with his girlfriend.

4. The amount of time he spends going out with his girlfriend is directly proportional to the amount of money he has in his wallet.

Use these assumptions to model Martin's grade in terms of the amount of money in his wallet. Don't forget to define the variables. If he wants a high grade, should he have more or less money in his wallet?

2.3.6 When a hot cup of coffee is set on a desk, it initially cools very quickly. As its temperature decreases, it does not cool as quickly. This suggests a proportionality relationship. Newton's law of cooling states that the rate at which a hot object (such as a hot cup of coffee) cools is proportional to the difference in the room temperature and the temperature of the object (assuming the room temperature stays constant). Define variables for the temperature of the object and the room temperature and set up a differential equation to model Newton's law of cooling (do not solve the equation).

2.3.7 A simple assumption for the spread of a contagious disease states that the rate at which the number of infected individuals changes is proportional to the product of the total number infected and the number not yet infected. Assume that initially one person carries the flu virus into a dorm with n residents. Let $x(t)$ represent the number of residents who are infected at time t. Set up a differential equation to model the spread of the flu through the dorm (do not solve the equation). (**Hint:** The total number of infected people is $[x + 1]$.)

2.4   Fitting Straight Lines Analytically

As we have seen, modeling with proportionality often requires us to fit a straight line to a set of data. Earlier, we used a graphical approach that can be rather subjective. In this section we look at different definitions of a "best-fit" line and find formulas for the slope and y-intercept of a best-fit line in terms of the x- and y-coordinates of the data points. This will give an objective approach to fitting a straight line.

The first step is to define criteria for a good-fitting line. Consider the two lines in Figure 2.12 fit to a set of three data points. We would say the dashed line fits the data and captures the trend of the data "better" than the solid line. What's the difference between these two lines? There are many ways to answer this question.

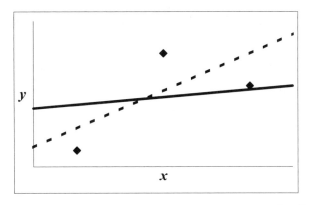

Figure 2.12

We see that the solid line is very close to the right-most point, but further from the other two points than the dashed line. We might say that the dashed line is "closer" to the data points in general than the solid line. The idea of minimizing the distance between the line and the points forms the basis of the definition of a best-fit line.

These distances (sometimes called errors) are illustrated in Figure 2.13 with the vertical lines. If the coordinates of the points are given by

$$(x_i, y_i) \text{ for } i = 1, 2, \ldots, n$$

and the line is described by the function $f(x) = mx + b$, then the values of the distances are

$$|y_i - f(x_i)| \text{ for } i = 1, 2, \ldots, n.$$

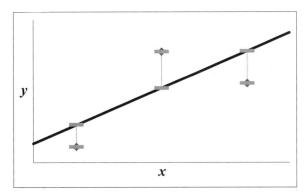

Figure 2.13

There are many ways to define how the best-fit line minimizes these distances. One way, called *Chebyshev's* criterion, is based on the idea that the best-fit line should make the largest of these distances as small as possible. In more technical terms, this criterion says that the function $f(x) = mx + b$ giving the best-fit line is the one that minimizes the number

$$\text{Maximum}\,\{|y_i - f(x_i)| \,:\, i = 1, 2, \ldots, n\}.$$

Our goal is to find formulas for the slope m and y-intercept b. Since we want to minimize something, we might think about using derivatives. Chebyshev's criterion makes logical sense, but it's not obvious how to take the derivative or use it to find simple formulas for m and b.

Another criterion is based on the idea that the best-fit line should minimize the sum of the distances. In mathematical notation, $f(x) = mx + b$ should minimize the number

$$\sum_{i=1}^{n} |y_i - f(x_i)|.$$

This criterion is also very logical, but the absolute values make the derivative difficult to calculate. To make the derivative simpler, we might consider getting rid of the absolute values in this summation and adding the criterion that the sum must be nonnegative. This, however, would make some of the terms in the summation positive and some negative. So, there might be some large positive values that cancel out some large negative values, resulting in a small sum but a poor-fitting line.

The most widely used criterion, called the *least-squares* criterion, uses squares rather than absolute values to make all the terms positive. In mathematical notation, this criterion says that $f(x) = mx + b$ should minimize the number

$$\sum_{i=1}^{n} (y_i - f(x_i))^2.$$

Taking the derivative of this expression looks relatively easy, so we use the least-squares criterion to find the desired formulas.

The goal is to find values of m and b in terms of the data points that minimize

$$S = \sum_{i=1}^{n} (y_i - f(x_i))^2 = \sum_{i=1}^{n} (y_i - mx_i - b)^2.$$

Since there are two variables (m and b), a necessary condition for optimality is that the partial derivatives with respect to each of these variables are zero. This gives the equations

$$\frac{\partial S}{\partial m} = \sum 2\,(y_i - mx_i - b)\,(-x_i) = -2\sum (y_i - mx_i - b)\,x_i = 0$$

$$\frac{\partial S}{\partial b} = \sum 2\,(y_i - mx_i - b)\,(-1) = -2\sum (y_i - mx_i - b) = 0$$

where all summations are from 1 to n. Rewriting these equations and solving for m and b yields the formulas

$$b = \frac{\sum x_i^2 \sum y_i - \sum x_i y_i \sum x_i}{n \sum x_i^2 - (\sum x_i)^2} \quad \text{and} \quad m = \frac{n \sum x_i y_i - \sum x_i \sum y_i}{n \sum x_i^2 - (\sum x_i)^2}. \tag{2.5}$$

Example 2.4.1 Least-Squares

We will use the least-squares criterion to find the best-fit line for a set of data in four different ways. The data is shown in columns **A** and **B** in Figure 2.14.

1. Rename a blank worksheet "**Least-Squares**" and format it as in Figure 2.14.

	A	B	C	D	E
1	x	y	f(x$_i$)	(y$_i$–f(x$_i$))²	Sum of Squares
2	0.8	2			
3	2.5	4.2			
4	3.5	3.5			
5	4.2	5.3			
6	5.8	4.5			
7	7.5	7.5			
8				Fitted Line	
9	m=	0.5	X	y	
10	b=	2	0	=B10	
11			8	=C11*B9+B10	

Figure 2.14

2. Highlight the data in the range **A2:B7** and create a graph as in Figure 2.15.

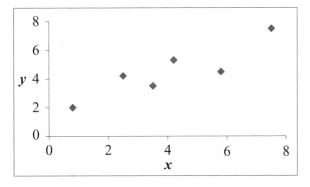

Figure 2.15

3. Highlight the data in the range **C10:D11** and drag it onto the graph. Add the cells as a **New Series** and format the graph so it looks like Figure 2.16.

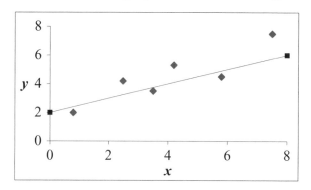

Figure 2.16

4. Add the formulas in Figure 2.17 to calculate the square of the distances and the sum of these squares. Copy the range **C2:D2** down to row 7.

	C	D	E
2	=$B$9*A2+$B$10	=(B2 – C2)^2	=SUM(D2:D7)

Figure 2.17

5. Now we use the **Solver** function to find the values of m and b that give the minimum total sum of squares. Select **Tools** → **Solver** ... . (**Note:** If you do not have the **Solver** ... option under **Tools**, select **Tools** → **Add-Ins** ... and check the box next to **Solver Add-in**. You may need your Excel installation CD to complete this.)

   (a) Next to **Set Target Cell:**, select **$E$2**.
   (b) Next to **Equal To:**, select **Min**.
   (c) Under **By Changing Cells:**, select **$B$9:$B$10**.
   (d) Click the **Solve** button.

   This gives $m = 0.694$ and $b = 1.689$.

6. Next we add formulas to analytically calculate the values of m and b using the formulas in equation (2.5). Add the formulas in Figure 2.18. Note that these are the same values of m and b as those found with **Solver** in Step 5.

7. Next we use a feature built into the graph to find the best-fit line. Right-click on the data in the chart and select **Add Trendline** ... . Under the **Type** tab, choose **Linear**. Under the **Options** tab, select **Display equation on chart**. Note that the slope and y-intercept of this line have the same values as m and b found in Step 5.

	F	G	H
1	**Sums**	**x**	**y**
2		=SUM(A2:A7)	=SUM(B2:B7)
3	n =	6	
4	m =	=(G3*J2-G2*H2)/(G3*I2-G2^2)	
5	b =	=(I2*H2-J2*G2)/(G3*I2-G2^2)	
6			
7	**Formulas**		
8	m =	=SLOPE(B2:B7,A2:A7)	
9	b =	=INTERCEPT(B2:B7,A2:A7)	

	I	J
1	x^2	xy
2	=SUMPRODUCT(A2:A7,A2:A7)	=SUMPRODUCT(A2:A7,B2:B7)

Figure 2.18

8. Lastly, Excel has built-in functions to calculate the values of the slope and the y-intercept. Add the formulas in Figure 2.19. Again, note that the values of m and b are the same as those found in Step 5.

	F	G
7	**Formulas**	
8	m =	=SLOPE(B2:B7,A2:A7)
9	b =	=INTERCEPT(B2:B7,A2:A7)

Figure 2.19

Exercises

2.4.1 Suppose a biologist records the number of pulses per second of the chirps of a cricket at different temperatures (in °F). The data collected is shown in Table 2.5.

Temperature	72	73	89	75	93	85	79	97	86	91
Pulses/s	16	16.2	21.2	16.5	20	18	16.75	19.25	18.25	18.5

Table 2.5

1. Fit a straight line to this data (where temperature is on the x-axis). How well does the model fit the data?

2. What is the slope of this line? What does the sign of the slope tell you about the relationship between pulses/s and temperature?

Year	1	2	3	4	5	6	7	8	9	10	11	12
Deaths	15	24	44	44	39	43	50	47	53	38	35	49
Year	13	14	15	16	17	18	19	20	21	22	23	24
Deaths	42	60	54	67	82	78	81	95	73	69	79	92

Table 2.6

2.4.2 Table 2.6 gives the number of manatee deaths in Florida believed to be caused by watercraft for the years 1983–2006 where Year 1 corresponds to 1983 (Florida Fish and Wildlife Conservation Commission, 2008). Fit a straight line to this data (where year is on the x-axis). How well does the model fit the data?

2.4.3 As we saw in Section 2.2, oftentimes we want to fit a straight line that goes through the origin to a set of data. This means we want the y-intercept of the straight line to be 0. So the equation of the line is simply $y = mx$. In this case, the least-squares criterion says that we want to minimize

$$S = \sum_{i=1}^{n} (y_i - mx_i)^2.$$

1. Take the derivative of this equation with respect to m, set it equal to 0, and solve for m to find a formula for m in terms of the x- and y-coordinates.

2. Implement this formula in Excel and use it to fit a straight line through the origin to the data in Table 2.1 on page 15. How does the slope of this line compare to the slope of the line found in Example 2.2.1?

3. This formula can be implemented by adding a trendline to the data and selecting **Set intercept =** under **Options**. Do this to the data and compare the slope of the trendline to the slope calculated by your formula. Are they indeed equal?

2.4.4 Generalize your result in 2.4.3. Suppose you have specified a value of b, say $b = b_0$, so that the model is now $y = mx + b_0$. In this case, the least-squares criterion says that we want to minimize

$$S = \sum_{i=1}^{n} [y_i - (mx_i + b_0)]^2.$$

1. Take the derivative of this equation with respect to m, set it equal to 0, and solve for m to find a formula for m in terms of the x- and y-coordinates and b_0.

2. Suppose you hang a bucket from a spring that stretches it 15 mm from its natural length. You then fill the bucket with different weights of sand and record the total distance stretched as recorded below.

Weight (N)	10	15	20	25	30	35
Distance (mm)	57.4	67.6	71.6	104.5	106.9	136.0

Because the spring was stretched before the first amount of sand was added, Hooke's law predicts that the relationship between the distance stretched, D, and the weight of sand, W, is

$$D = mW + 15.$$

Create a graph of the data to test if this prediction is reasonable. If it is, use your formula from part (1) to estimate the value of m.

3. We could also estimate the value of m by transforming the data by subtracting 15 from each distance measurement, and then fitting a straight line through the origin to the transformed data. The slope of this line is m. Do this and compare the value of m to that found in part (2).

2.4.5 Sports Manufacturing Inc. manufactures footballs, basketballs, and soccer balls. Each week the company manufactures a different type of ball in varying quantities. Manufacturing costs fall into two different categories: start-up and unit. Start-up costs are costs necessary to begin production of a particular product (e.g., retool machinery). Unit costs ($/unit) are the costs associated with manufacturing individual units (e.g., labor, materials).

The table below shows data for 15 weeks of production. The "Total Cost" column gives the total cost to produce the given number of units of that type of ball in a week. Your goal is to model the total cost, estimate the start-up and unit costs for each product, and implement the model.

Footballs		Basketballs		Soccer Balls	
Units	Total Cost	Units	Total Cost	Units	Total Cost
2222	3125	962	1520	2481	4300
2263	3250	2246	2850	1825	3190
1267	1955	2430	2990	2238	3930
2177	3120	1395	1920	949	1890
2266	3090	2405	2750	1250	2350

1. Define variables and create a model for the total cost for producing a given number of units of each product in terms of the start-up and unit costs. List the assumptions you make.

2. Estimate the start-up and unit costs for each product.

3. Create a spreadsheet in which a user can easily input the number of units of each product that are planned for production in a week. Have the spreadsheet automatically calculate the estimated total production cost for each product. Make sure the spreadsheet is logical and easy to use.

2.5   Geometric Similarity

Shapes such as circles and rectangles are easy to work with. We can calculate the area and volume of objects with these shapes using very simple formulas. Real-world objects rarely come in these simple forms. This necessitates some simplifying assumptions. Geometric similarity is one such assumption.

Definition 2.5.1 Two objects are *geometrically similar* if the following two conditions are met:

1. There is a one-to-one correspondence between points of the objects (i.e., the two objects have the same "shape").

2. The ratio of distances between corresponding points is the same for all pairs of points.

In simpler terms, two objects are geometrically similar if one is a scaled-up or scaled-down version of the other.

What does geometric similarity allow us to do? Let's start with a very simple example. Consider a rectangle of length 3 cm and height 2 cm. Its area is 6 cm². Note that this area is proportional to the square of the length of the rectangle since

$$6 = \frac{6}{3^2} \left(3^2\right) = \frac{2}{3} \left(3^2\right).$$

Now consider another rectangle of length 5 cm and height 4 cm. Its area is 20 cm², which is again proportional to the square of the length since

$$20 = \frac{20}{5^2} \left(5^2\right) = \frac{4}{5} \left(5^2\right).$$

Consider a third rectangle of length 6 cm and height 4 cm. This rectangle is a scaled-up version of the first rectangle (its dimensions are simple two times the dimensions of the first one). In other words, these two rectangles are geometrically similar. The new rectangle's area is 24 cm², which is again proportional to the square of the length since

$$24 = \frac{24}{6^2} \left(6^2\right) = \frac{2}{3} \left(6^2\right).$$

Notice that the constants of proportionality are the same for these two geometrically similar rectangles. The second rectangle is not geometrically similar to the other two rectangles, and its constant of proportionality is not the same as the others.

Let's generalize this example. Suppose we have a rectangle of length $3k$ cm and width $2k$ cm where k is some positive number. This rectangle is geometrically similar to the first rectangle. Its area is $6k^2$ cm², which is proportional to the square of the length since

$$6k^2 = \frac{6k^2}{(3k)^2} \left(3k\right)^2 = \frac{2}{3} \left(3k\right)^2.$$

Again note that the constant of proportionality is the same as for the other geometrically similar rectangles. What does this mean? Suppose we have a bag of rectangles, each of length $3k$ cm and width $2k$ cm where $k > 0$ is different for each rectangle. If we reached into the bag and pulled out one rectangle and wanted to know its area, we wouldn't need to measure *both* the length and the height. We could simply measure the length, square it, and take it times 2/3. This simplifies the process of finding the area.

The length is an example of a *characteristic dimension*. A characteristic dimension is simply a dimension of the object that is easy to measure. We could have chosen height as the characteristic dimension and done the same analysis as above, but the constant of proportionality would have been different.

This generalization illustrates the first important property of geometrically similar objects.

Theorem 2.5.1 *Suppose H is a set of geometrically similar objects. Let S denote the surface area of an object and l denote a characteristic dimension. Then*

$$S \alpha l^2,$$

and the constant of proportionality is the same for every object in H.

Notice that no certain shape, or dimension, of the objects is assumed in Theorem 2.5.1. This property allows us to simplify the modeling of the area of complex shapes.

Example 2.5.1   Wool from a Sheep

A shepherd wants to predict the volume of wool he will get from a sheep, V, in terms of the girth of the sheep (the distance around the fattest part of the belly). The volume is the thickness of the wool times the area from which it is shaved. This area is not a simple shape, so we will use geometric similarity to simplify the model.

Consider the following assumptions:

1. The thickness of the wool is the same for every sheep.
2. The area on each sheep from which the wool is shaved is geometrically similar.

The first assumption allows us to model

$$V \alpha S \tag{2.6}$$

where S is the area from which the wool is shaved. The second assumption allows us to model

$$S \alpha l^2 \tag{2.7}$$

where l is some characteristic dimension. We will choose the girth (the distance around the belly of the sheep) to be this dimension. Combining (2.6) and (2.7) using transitivity, we get

$$V \alpha l^2. \tag{2.8}$$

To find the constant of proportionality and test the assumptions, we would need to collect data of volume and girth.

Now let's consider a three-dimensional object. Specifically, consider a rectangular box with width 4 cm, height 3 cm, and depth 2 cm. Its volume is 24 cm^3, which is proportional to the height cubed since

$$24 = \frac{24}{3^3} \left(3^3\right) = \frac{8}{9} \left(3^3\right).$$

Consider a geometrically similar box with width $4k$ cm, height $3k$ cm, and depth $2k$ cm where $k > 0$. Its volume is $24k^3$ cm^3, which again is proportional to the height cubed since

$$24k^3 = \frac{24k^3}{(3k)^3} \left(3k\right)^3 = \frac{8}{9} \left(3k\right)^3.$$

Note that the constant of proportionality is the same. This generalization illustrates the second important property of geometrically similar objects.

Theorem 2.5.2 *Suppose H is a set of geometrically similar objects. Let V denote the volume of an object and l denote a characteristic dimension. Then*

$$V \alpha l^3,$$

and the constant of proportionality is the same for every object in H.

As in the first property, no special shape of the objects is assumed. This property allows us to relate the volume of an object to some characteristic dimension, and combining this with the first property we can relate volume to surface area.

Example 2.5.2   Surface Area of a Potato

Let's suppose we want to fix a large batch of the recipe "Crispy Potato Skins" for an appetizer at our Super Bowl party. This recipe requires only the skin from a potato, so when we buy the potatoes, we want to get the maximum surface area for our money.

At the supermarket we have the choice of several different sizes of potatoes. We need to decide whether to buy several small potatoes or a few large ones (we are assuming that we can choose individual potatoes).   Let's restrict ourselves to the following problem:

Should we buy eight small baking potatoes weighing 0.25 lb each, or four large baking potatoes weighing 0.5 lb each?

To answer this question, we want to relate the surface area of a potato, A, to its weight, W. Consider the following assumptions:

1. Potatoes have a constant density.
2. Potatoes are geometrically similar.

The first assumption seems very accurate. The veracity of the second is arguable. However, potatoes have an irregular shape, so we need some sort of simplifying assumption to model their surface. Similarity is a *reasonable* assumption.

Now, weight = density × volume, so the first assumption allows us to model

$$W \alpha V \qquad (2.9)$$

where W is the weight and V is the volume. The second assumption allows us to model

$$V \alpha l^3 \qquad (2.10)$$

where l is *any* characteristic dimension (such as length). Combining (2.9) and (2.10) we get

$$W \alpha l^3. \qquad (2.11)$$

The second assumption also allows us to model

$$A \alpha l^2 \qquad (2.12)$$

where l is the same characteristic dimension used in (2.11). Rewriting (2.11) and combining it with (2.12), we get

$$l \alpha W^{1/3} \rightarrow A \alpha \left(W^{1/3}\right)^2 = W^{2/3}. \qquad (2.13)$$

Since potatoes are sold by the pound, each choice in the original problem will cost the same amount, so we want the choice with the largest surface area. If A_S and A_L represent the total surface area of the small and large potatoes, respectively, then (2.13) gives

$$A_S = 8k \left(0.25\right)^{2/3} \text{ and } A_L = 4k \left(0.5\right)^{2/3}$$

where k is some constant (note k is the same for both A_S and A_L). Thus

$$\frac{A_S}{A_L} = \frac{8k\,(0.25)^{2/3}}{4k\,(0.5)^{2/3}} \approx 1.26 \quad \Rightarrow \quad A_S \approx 1.26A_L.$$

Therefore, the surface area of the small potatoes is approximately 26% greater than the surface area of the larger potatoes. We should buy the smaller potatoes.

As seen in Example 2.5.2, even if we don't know the constant of proportionality, we can still use a proportionality model to come to a conclusion. The next example also illustrates this. In addition, it illustrates the point that we can use Theorems 2.5.1 and 2.5.2 to relate the area or volume of a *portion* of an object to some characteristic dimension of the whole object.

Example 2.5.3   Agility
A sport such as gymnastics is said to require great agility. Anyone who has watched gymnastics has noted that there are very few tall gymnasts. Why is this? To answer this question, we construct a model that relates the height of a person, h, to his or her agility (adapted from Giordano, Weir, and Fox, 2003, 95–96).

First we need to define agility. It seems reasonable that the stronger you are, the more agile you are (up to a point). Also, it seems reasonable that the heavier you are, the less agile you are. Thus, the following mathematical "definition" of agility seems reasonable:

$$\text{agility} \; \alpha \; \frac{\text{strength}}{\text{weight}} \tag{2.14}$$

Next we need to relate strength and weight to height. Consider the following assumptions:

1. The strength of a muscle is proportional to its cross-sectional area.
2. All gymnasts are geometrically similar.
3. All gymnasts have the same body density.

The first assumption is supported by physiological arguments. The second two assumptions are arguable, but reasonable. Using assumptions (1) and (2), we can relate strength to the characteristic dimension of height by

$$\text{strength} \; \alpha \; \text{cross-sectional area} \; \alpha \; h^2. \tag{2.15}$$

Using assumptions (2) and (3), we can relate weight to height by

$$\text{weight} \; \alpha \; h^3. \tag{2.16}$$

So, combining (2.14), (2.15), and (2.16), we get

$$\text{agility} \, \alpha \, \frac{\text{strength}}{\text{weight}} \, \alpha \, \frac{h^2}{h^3} \, \alpha \, \frac{1}{h}.$$

Thus agility is inversely proportional to height. This means that as height increases, agility decreases, explaining why gymnasts are typically very short.

Exercises

2.5.1 Answer the following questions related to geometric similarity and proportionality:

1. If water bottles are geometrically similar, how much more water will a bottle that is 30 cm tall hold than one that is 10 cm tall?

2. If air resistance is proportional to the surface area of a falling object at a given velocity, how much more air resistance will a sphere of diamater 5 cm encounter than one with a diameter of 1 cm?

3. If gymnasts have a constant density, then weight is proportional to volume. If we further assume that they are geometrically similar, how much less would a gymnast who is 5 ft tall weigh than one who is 5.5 ft tall?

4. If hearts are geometrically similar and the volume of blood pumped in one beat is proportional to the volume of the heart, how much more blood will a heart that is 4 cm wide pump in one beat than a heart that is 1 cm wide?

5. If the amount of heat lost by a submarine over a unit of time is proportional to its surface area, how much more heat will a submarine that is 50 ft long lose over a given period of time than a scale model that is 10 ft long?

6. If objects are geometrically similar and have a constant density, we saw in Example 2.5.2 that $A \, \alpha \, W^{2/3}$ where A = surface area and W = weight. If the weight of one such object is five times the weight of another, how much larger is the surface area?

7. Suppose that an ice cube melts so that at any point in time, the remaining cube is geometrically similar to the initial cube (i.e., before it started melting). At one point in time, the length is half the initial length. What fraction of the initial volume has melted?

2.5.2 In Example 2.5.2, we assumed that potatoes are geometrically similar. Describe how you might collect data to test the reasonableness of this assumption and the validity of the model derived in this example.

2.5.3 Suppose it takes 0.75 oz of sunscreen to cover all exposed areas of a 3-ft-tall child. Estimate the amount of sunscreen it would take to cover all exposed areas of a 6-ft-tall man. List all assumptions you make.

2.5.4 Consider a set of geometrically similar objects.

1. Derive a model for the ratio $\dfrac{\text{surface area}}{\text{length}}$ in terms of length. (**Hint:** Use length as the characteristic dimension and model surface area in terms of length.)

2. Derive a model for the ratio $\dfrac{\text{volume}}{\text{length}}$ in terms of length.

2.5.5 Ace Manufacturing Company owns a warehouse in the shape of a rectangular box that measures 100 ft wide by 200 ft long by 10 ft high. It has a furnace with an output of 500,000 BTUs. Ace wants to build a larger warehouse measuring 200 ft by 400 ft by 20 ft and is trying to determine what size of furnace needs to be installed.

1. If the company assumes that the size of the furnace is proportional to the volume of the building, what size furnace should be installed?

2. If the company assumes that the size of the furnace is proportional to the total surface area of the building (including the floor), what size furnace should be installed?

3. If the company assumes that the size of the furnace is proportional to the surface area of the walls and roof *only* (meaning the floor is not a factor), what size furnace should be installed?

2.5.6 A rowing shell is a slim, needle-like boat built for speed that is powered by one, two, four, or eight rowers. We want to build a model that predicts the boat's speed, v, in terms of the number of rowers, r. Table 2.7 contains data on boat length and winning race times at four world championships (reprinted with permission from *Science News*, copyright 2008).

Rowers (r)	Length (l)	Time for 2000 m (min)			
		I	II	III	IV
8	18.28	5.87	5.92	5.82	5.73
4	11.75	6.33	6.42	6.48	6.13
2	9.76	6.87	6.92	6.95	6.77
1	7.93	7.16	7.25	7.28	7.17

Table 2.7

Consider the following assumptions:

1. $l \alpha r^{1/3}$ where l is the length of the boat.

2. The only drag slowing down the boat is from the water, which is proportional to Sv^2 where S is the wetted surface area of the boat.

3. The boats travel at a constant speed. This means that the force applied by the rowers equals the force of drag.

4. All boats are geometrically similar with regard to S.

5. The power available to the boat is proportional to r. (Power = force × speed where force is the force applied by the rowers.)

Is the first assumption reasonable? Build a model that predicts v in terms of r and test it with the given data. Here are a few suggestions:

1. Start with the relationship power = force × speed and substitute the proportionality relationships from the other assumptions.

2. Remember, velocity = distance/time.

3. Use the average time for the 2000 m race to calculate the velocity.

2.5.7 Consider a raindrop falling from a cloud. Ignoring any effects of wind, it is influenced by two forces: weight due to gravity and air resistance. At some point, the raindrop reaches a terminal velocity, v_t, where these two forces are equal. Consider the following assumptions:

1. All raindrops are geometrically similar.

2. All raindrops have the same density.

3. Air resistance is proportional to the product of the raindrop's surface area and the square of its speed.

Use these assumptions to model the terminal velocity of a raindrop in terms of its weight.

For Further Reading

- For more examples of modeling with proportionality, see F. R. Giordano, M. D. Weir, and W. P. Fox. 2003. *A first course in mathematical modeling.* 3rd ed. Pacific Groves, CA: Thomson Brooks/Cole, 95–96.

- For examples of modeling biological systems with proportionality, see A. J. Clark. 1927. *Comparative physiology of the heart.* Cambridge, UK: Cambridge University Press.

References

Florida Fish and Wildlife Conservation Commission. Marine Mammal Pathobiology Laboratory. 2009. *http://research.myfwc.com/features/view_article.asp?id=12084*

Giordano, F. R., M. D. Weir, and W. P. Fox. 2003. *A first course in mathematical modeling.* 3rd ed. Pacific Groves, CA: Thomson Brooks/Cole.

Science News. 2008. *http://www.sciencenews.org*

CHAPTER 3

Empirical Modeling

Chapter Objectives

- Fit various types of models to a set of data to predict values
- Use the coefficient of determination to assess how well a model fits a set of data
- Fit polynomial models to data
- Introduce multiple regression
- Introduce spline models

3.1   Introduction

In previous chapters we used theory of one form or another to construct models and then used data to determine the value of constants within the model. This process is often called *model fitting*. The model never fit the data perfectly, but we were willing to accept some error because the model helps *explain* the behavior of the system. These models are often called *analytical models*.

In this chapter we build models guided solely by data. We do not even attempt to use theory to explain behavior. Rather, we find a model that captures the trend of the data and use it to *predict* values rather than explain the behavior. These models are often called *empirical models*. Many of the topics in this chapter are closely related to the field of statistics, and particularly the topic of regression.

3.2   Linearizable Models

Linearizable models are those that can be fit to a set of data by making an appropriate transformation and then fitting a linear model to the transformed data. Listed below are the three most basic types of linearizable models:

Logarithmic	Power	Exponential
$y = a + b \ln(x)$	$y = ax^b$	$y = ae^{bx}$

The variable x is called the *predictor* variable while the variable y is called the *response* variable. Graphs of these different types of models are shown in Figure 3.1. Note the "shape" of the different graphs. Each one of these different functions increases as x increases, but they increase at different rates. The logarithmic function and the power functions with exponents less than 1 increase much more slowly than the other types of functions. They almost appear to "level off," whereas the other types grow very quickly. Recognizing the shape of the different graphs will help us to select an appropriate type of model.

To illustrate how to use these models, consider the data in Table 3.1, which gives the number of persons per physician and male life expectancy (in years) for various countries around the world (*World Almanac Book of Facts*, 1992). Our goal is to predict life expectancy in terms of the number of persons per physician.

Obviously there are many factors involved with life expectancy; the number of persons per physician is only one of them. It seems reasonable to believe that as the number of persons per physician increases (meaning fewer doctors per person), life expectancy decreases because people would not have as easy access to health care. We do not claim

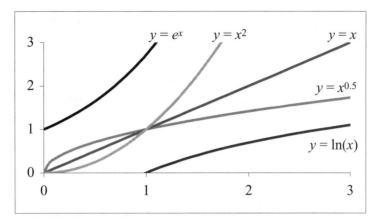

Figure 3.1

Country	Persons/Physician P	Male Life Expectancy L
Spain	275	74
United States	410	72
Canada	467	73
Romania	559	67
China	643	68
Taiwan	1010	70
Mexico	1037	67
South Korea	1216	66
India	2471	57
Morocco	4873	62
Bangladesh	6166	54
Kenya	7174	59

Table 3.1

that the number of persons per physician *causes* life expectancy, but there is probably a relationship between the two variables. It is not at all clear how the variables of persons per physician and life expectancy are related theoretically, so we will not even attempt to construct an analytical model. We will construct an empirical model by fitting various linearizable models to this data and analyzing how well each one fits.

When constructing any type of empirical model, the first step is plot y versus x (L versus P in this case) and look at the "shape." The second step is to select an appropriate type of model and fit it to the data.

A graph of the data is shown in Figure 3.2. Notice that as the number of persons per physician increases, the life expectancy decreases, agreeing with our intuition. Also note that the points seems to form a curve that initially decreases rapidly, but then levels off. This suggests that a logarithmic or power model might be appropriate.

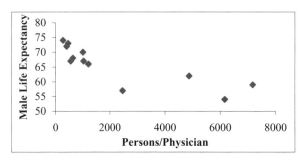

Figure 3.2

Example 3.2.1   Logarithmic Model

We will fit a curve of the form $L = a + b \ln(P)$ to the data by graphing L versus $\ln(P)$ and fitting a straight line. The value of b is the slope of this line and the value of a is the y-intercept.

1. Name a blank worksheet "**ln**" and format it as in Figure 3.3. Enter the rest of the data from Table 3.1 in columns **A** and **C**. Copy cell **B2** down to row 13.

	A	B	C
1	P	ln(P)	L
2	275	=LN(A2)	74

Figure 3.3

2. Create a graph of L versus $\ln(P)$, add a linear trendline, and display the equation of the line as in Figure 3.4. Using the slope and y-intercept of this line, we get our model: $L = 103.4 - 5.2876 \ln(P)$.

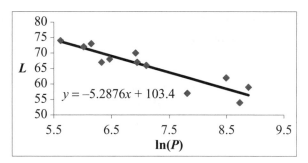

Figure 3.4

3. Next we need to compare the model to the original data. Add the formula in Figure 3.5 and copy cell **D2** down to row 13.

	D
1	**Predicted**
2	=103.4 − 5.2876*LN(A2)

Figure 3.5

4. Create a graph to compare the observed values of L to the predicted ones as in Figure 3.6. Notice that the two sets of values are fairly close together, indicating that we have a good model.

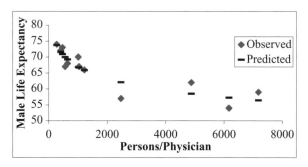

Figure 3.6

5. To further analyze how well the model fits the data, for each data point define

$$\text{Residual} = (\text{Observed value}) - (\text{Predicted value}).$$

Note that a positive residual means that the predicted value is less than the observed value. A negative value means that the predicted value is greater than the observed value. To calculate the residual for each data point, add the formula in Figure 3.7 and copy cell **E2** down to row 13.

	E
1	**Residual**
2	= C2–D2

Figure 3.7

6. Create a graph of Residual versus P as in Figure 3.8. Note that roughly half the residuals are positive and half are negative. This indicates that the model does not tend to overpredict or underpredict the values of L. Also note that the magnitudes of the residuals (the absolute values) are all relatively small, less than 6, and that there is no "pattern" to the residuals. These three observations indicate that this model fits relatively well.

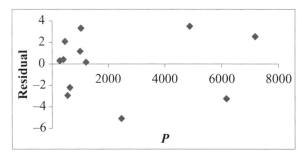

Figure 3.8

Example 3.2.2   Power Model

We will fit a curve of the form $L = aP^b$ to the data. To find the values of a and b, we take the natural logarithm of both sides of the model to get

$$\ln L = \ln(aP^b) = \ln a + \ln P^b = \ln a + b \ln P.$$

Thus a straight line fit to the graph of $\ln L$ versus $\ln P$ will have a slope of b and a y-intercept of $\ln a$.

1. Name a blank worksheet "**Power**" and format it as in Figure 3.9. Copy the rest of the data from worksheet **ln** into columns **A** and **B**. Copy the range **C2:D2** down to row 13.

	A	B	C	D
1	P	L	ln(P)	ln(L)
2	275	74	= LN(A2)	= LN(B2)

Figure 3.9

2. Graph ln L versus ln P, fit a straight line, and display the equation on the graph as in Figure 3.10.

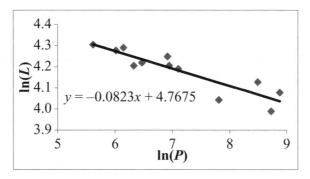

Figure 3.10

The equation of the line is $y = -0.0823x + 4.7675$, so $b = -0.0823$ and

$$\ln a = 4.7675 \quad \Rightarrow \quad a = e^{4.7675} = 117.62.$$

Therefore, the model is $L = 117.62 P^{-0.0823}$.

3. Add the formulas in Figure 3.11 to calculate the predicted values and the residuals. Copy row 2 down to row 13.

	E	F
1	Predicted	Residual
2	=117.62*A2^−0.0823	= B2−E2

Figure 3.11

4. Create a graph of Residual versus P as shown in Figure 3.12. Note that again roughly half of the residuals are positive and half are negative, they all have magnitudes less than 5, and there is no pattern. This indicates a good model.

Example 3.2.3   Exponential Model

The plot of the data does not resemble the graph of an exponential model, so this type of model may not be the best. However, we will fit one to the data to illustrate the process.

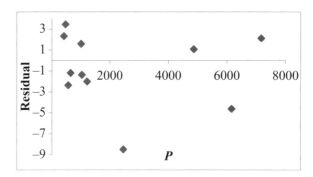

Figure 3.12

The exponential model has the form $L = ae^{bP}$. To find the values of a and b, we take the natural logarithm of both sides of the model to get

$$\ln(L) = \ln(ae^{bP}) = \ln a + \ln(e^{bP}) = \ln a + bP.$$

Thus a straight line fit to the graph of $\ln L$ versus P will have a slope of b and a y-intercept of $\ln a$.

1. Name a blank worksheet "**Exponential**" and format it as in Figure 3.13. Copy the rest of the data from the worksheet **Power** and copy cell **C2** down to row 13.

	A	B	C
1	P	L	ln(L)
2	275	74	=LN(B2)

Figure 3.13

2. Create a graph of $\ln L$ versus P and fit a straight line to the data as in Figure 3.14. Display the equation on the chart. Notice that this line does not fit the data as well as with the other two models. This is an indication that an exponential model does not fit the data as well as the others.

The equation of this line is $y = -0.00003x + 4.2561$, so $b = -0.00003$ and

$$\ln a = 4.2561 \quad \Rightarrow \quad a = e^{4.2561} = 70.53.$$

Therefore, the model is $L = 70.53\, e^{-0.00003P}$.

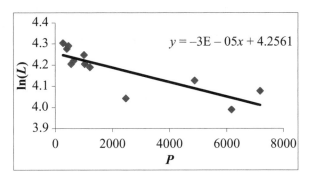

Figure 3.14

3. Add the formulas in Figure 3.15 to calculate the predicted values and the residuals. Copy row 2 down to row 16.

	D	E
1	**Predicted**	**Residual**
2	=70.53*EXP(−0.00003*A2)	=B2−D2

Figure 3.15

4. Create a graph of Residual versus P as shown in Figure 3.16. Notice that all the residuals are at least 1 in magnitude and that one is almost -9. This indicates that the model does not predict any of the values of L very accurately. Therefore, this is not the best fitting model, as previously suspected.

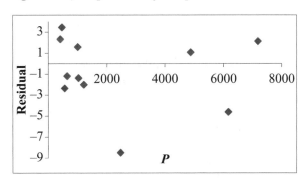

Figure 3.16

Example 3.2.4   Trendlines

Excel will automatically calculate these different models for us. Create a graph of L versus P, add a trendline, select the type of model you want under **Type**, and display the equation on the chart. The results are shown in Figure 3.17. Note that these are exactly the same models we derived. Also note that the graphs of the logarithmic and power models capture the trend of the data very well while the exponential model does not. This confirms our conclusions based on the graphs of the residuals.

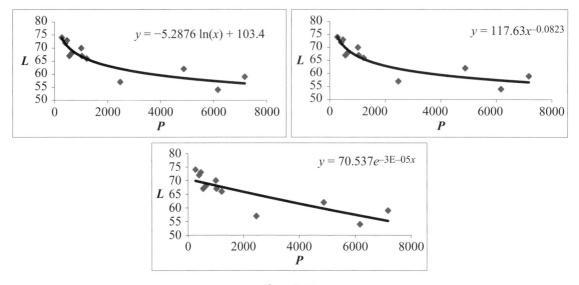

Figure 3.17

Out of the three models, the logarithmic and power models are the "best" based on an analysis of the graph of the models and of the residuals. In the next section we look at more analytical techniques for measuring how well a model fits a set of data.

Now that we have two good-fitting models for the data, what can we do with them? There are at least two uses. First of all, the graphs of the models help us to see the trend of the data. The graphs decrease from left to right, helping us to illustrate the point that as the number of persons per physician increases, life expectancy decreases. The plot of the data also shows this, but a curve helps exemplify the relationship.

Second of all, we can use the models to *predict* life expectancy if we know the number of persons per doctor. For instance, suppose a country has 3500 persons per physician. The logarithmic model predicts that life expectancy is

$$L = 103.4 - 5.2876 \ln{(3500)} \approx 60.25 \text{ years}$$

while the power model gives

$$L = 117.63 \left(3500\right)^{-0.0823} \approx 60.10 \text{ years}.$$

We certainly could plug $P = 3500$ into the exponential model and get

$$L = 70.537e^{-0.00003(3500)} \approx 63.51 \text{ years.}$$

However, we saw that the exponential model did not fit the data very well, so it would not be appropriate to use it for making predictions. This illustrates the first caution when using empirical models: *If a model does not fit a set of data, do not use it for making predictions.*

Now suppose a country has 100 persons per physician. We could plug $P = 100$ into the logarithmic model and get

$$L = 103.4 - 5.2876 \ln(100) \approx 79.04 \text{ years.}$$

However, note that $P = 100$ is outside the range of the original data. We do not know the trend of the data for values of P less than 275. It could change or stay the same; we simply do not know. Therefore, it would be inappropriate to use any of these models to predict values of L for P less than 275. The value of 79.04 years may or may not be accurate, so we should not report this "prediction." This illustrates the second caution when using empirical models: *Only use values of the predictor variable that are within the range of the original set of data.*

Exercises

3.2.1 For each set of data below, fit a model of the given form by transforming the data appropriately and fitting a straight line to the transformed data. Graph the residuals and analyze how well the model fits the data.

1. Model: $y = ax^2 + b$

x	1	2	3	4	5	6
y	16.3	23.1	37.4	46.9	58.7	91.0

2. Model: $y = a \sin(x) + b$

x	1	2	3	4	5	6
y	1.34	1.61	-0.98	-3.80	-4.55	-2.30

3. Model: $y = a\frac{x^2+1}{\ln(x)} + b$

x	2	3	4	5	6	7
y	3.30	5.63	9.52	14.31	19.84	26.061

3.2.2 Consider the data below:

x	0	2	4	6	8	10
y	3.00	3.06	3.12	3.19	3.25	3.32

1. Fit a linear model to the data (a model of the form $y = mx + b$). How well does the model appear to fit the data? Create a graph of the residuals. What do you notice?

2. Fit an exponential model to the data. How well does the model appear to fit the data? Calculate the residuals. What does this tell you about how well the model fits the data?

3.2.3 For each of the data sets below, determine which model — exponential, power, or logarithmic — best fits the data.

1.
x	1	2	3	4	5	6
y	1.66	2.41	6.04	9.89	17.31	31.54

2.
x	1	2	3	4	5	6
y	2.68	5.61	8.71	9.83	11.16	11.03

3.3   Coefficient of Determination

The coefficient of determination, denoted R^2, is a numerical measure of how well a line fits a set of data. To illustrate the fundamental concepts, we will generate some hypothetical data according to the relationship $y = 3 + 2x$.

1. Rename a blank worksheet "**R2**" and format it as in Figure 3.18. Copy row 3 down to row 11 to generate 10 data points.

	A	B
1	**x**	**y**
2	0	=3+2*A2
3	=A2+1	=3+2*A3

Figure 3.18

2. Create a graph of the data and add a trendline as in Figure 3.19. Note that the line goes through each data point and the equation of this line (also called the *regression equation*) is exactly what was used to generate the data.

One purpose of a fitting a line to data is to use it for predicting the value of y when x is known. If we did not have a graph of the data or the regression equation and we were given

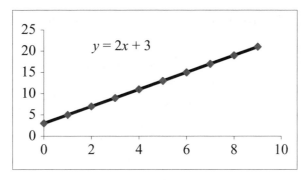

Figure 3.19

a value of x, the best guess as to the corresponding value of y would be the mean of the y-values. This mean is denoted by $\bar{y}$ and equals 12 in this case.

We test this simple prediction strategy by examining the "error" it would cause for the given data points. This error, or deviation, $y - \bar{y}$, is illustrated in Figure 3.20.

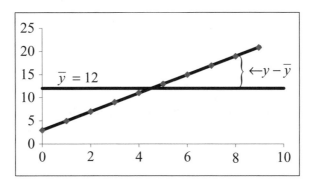

Figure 3.20

If a y-value predicted by the regression equation is denoted $\hat{y}$, we measure how well a regression equation fits the data by comparing the deviation to the difference between $\hat{y}$ and $\bar{y}$, $\hat{y} - \bar{y}$. In this case, the regression line goes through each data point, so $\hat{y} = y$. Therefore,

$$\hat{y} - \bar{y} = y - \bar{y} \quad \Rightarrow \quad \frac{\hat{y} - \bar{y}}{y - \bar{y}} = 1.$$

Thus we give this regression equation an R^2 value of 1. This value is often interpreted by saying that the regression equation "explains" 100% of the deviation. The definition of the coefficient of determination is based on this idea of comparing the deviation to the difference between $\hat{y}$ and $\bar{y}$ to measure the percentage of deviation "explained" by the regression equation.

This set of data is highly idealized because it was generated exactly according to the linear relationship $y = 3 + 2x$. In reality, data never conforms to an exact relationship like this. Real data with a linear relationship satisfies an equation of the form

$$y = \beta_0 + \beta_1 x + \varepsilon$$

where ε is some "noise." This noise may be due to measurement error, sampling variation, or some other random event outside of our control.

The relation $y = \beta_0 + \beta_1 x$ is called the "true" linear trend of the population while the regression equation has the generic form $\hat{y} = \hat{\beta}_0 + \hat{\beta}_1 x$. The "hats" indicate that the parameters $\hat{\beta}_0$ and $\hat{\beta}_1$, and the values of y, are estimates of the population values based on sample data.

Example 3.3.1   Including Noise

To generate data with some noise, we modify the worksheet **R2** as in Figure 3.21. Copy cell **B2** down to row 11. Here our "noise" is given by NORMINV(RAND(),0,2), which is a normally distributed pseudorandom variable with mean 0 and standard deviation 2.

	B
1	y
2	=3+2*A2+NORMINV(RAND(),0,2)

Figure 3.21

The graph of this noisy data should resemble Figure 3.22. Note that your graph will probably look different due to the random noise. Also note that the regression equation is *not* $y = 3 + 2x$.

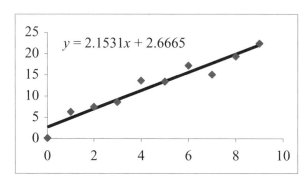

Figure 3.22

To define the R^2 value for a line fit to noisy data, we introduce three different types of deviation: *explained*, *unexplained*, and *total* deviation. Figure 3.23 illustrates these terms. We see from the figure that

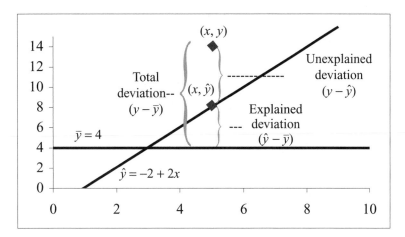

Figure 3.23

$$
\begin{array}{ccccc}
\text{(total deviation)} & = & \text{(explained deviation)} & + & \text{(unexplained deviation)} \\
(y - \bar{y}) & = & (\hat{y} - \bar{y}) & + & (y - \hat{y}).
\end{array}
$$

If we square these deviations and add them together for all data points, we get amounts of *variation* (measures of the total deviation). The total variation is called the *total sum of squares* and is given by

$$SS_{Tot} = \sum (y_i - \bar{y})^2.$$

The explained variation is called the *regression sum of squares* and is given by

$$SS_{Reg} = \sum (\hat{y}_i - \bar{y})^2.$$

The unexplained variation is called the *residual sum of squares* and is given by

$$SS_{Res} = \sum (y_i - \hat{y}_i)^2.$$

Analogous to the relationship between the deviations we get

$$
\begin{array}{ccccc}
\text{(total variation)} & = & \text{(explained variation)} & + & \text{(unexplained variation)} \\
\sum (y_i - \bar{y})^2 & = & \sum (\hat{y}_i - \bar{y})^2 & + & \sum (y_i - \hat{y}_i)^2
\end{array}
$$

Thus proving this relationship is true is not a trivial matter. Rewriting this equation we get

$$\text{(explained variation)} = \text{(total variation)} - \text{(unexplained variation)}$$

$$\begin{aligned} \sum(\hat{y}_i - \bar{y})^2 &= \sum(y_i - \bar{y})^2 &- \sum(y_i - \hat{y}_i)^2 \\ SS_{Reg} &= SS_{Tot} &- SS_{Res}. \end{aligned}$$

Therefore, the "percentage" of the total variation explained by the regression equation is

$$R^2 = \frac{SS_{Tot} - SS_{Res}}{SS_{Tot}}. \tag{3.1}$$

This is the definition of the coefficient of determination. The closer R^2 is to 1, the better the line fits the data. An R^2 value close to 0 indicates a very poor-fitting line.

Example 3.3.2   Calculating R^2

To use equation (3.1) to calculate R^2 for the regression line fit to the noisy data, follow these steps:

1. Modify the worksheet **R2** as in Figure 3.24.

	A	B
14	**Average=**	=AVERAGE(B2:B11)
15	**Slope=**	=SLOPE(B2:B11,A2:A11)
16	**y-int=**	=INTERCEPT(B2:B11,A2:A11)

Figure 3.24

2. Add the formulas in Figure 3.25 and copy row 2 down to row 11.

	C	D	E
1	**Predicted**	**SS$_{Tot}$**	**SS$_{Res}$**
2	=A2*$B$15+$B$16	=(B2–$B$14)^2	=(B2-C2)^2

Figure 3.25

3. Add the formulas in Figure 3.26.

	C	D	E
13	**Sum =**	= SUM(D2:D11)	= SUM(E2:E11)
14		**R²=**	= (D13-E13)/D13

Figure 3.26

4. Add a linear trendline to the graph of the data. Under **Options**, select **Display R-squared value on chart**. Note that this R^2 value is equal to what we calculated.

5. Press the **F9** key several times. Each time you press it, the random numbers in the noise are regenerated and a new set of data is created. Your R^2 value in cell **E14** should equal the one automatically generated by Excel each time.

   Notice that the R^2 value is not exactly 1. That is because our data has some noise, so the underlying linear relationship $y = 3 + 2x$, or any other linear relationship, does not account for all of the variation from the mean. The R^2 value is, however, very close to 1. This indicates that the regression line does fit the data very well.

6. To add more "noise" to the data, modify the formulas in the worksheet **R2** as in Figure 3.27 and copy cell **B2** down to row 11. Now the noise is normally distributed with mean 0 and standard deviation 4, so it is more "spread out" than before.

	B
1	y
2	=3+2*A2+NORMINV(RAND(),0,4)

Figure 3.27

Notice that the R^2 value is less than before. The straight line cannot account for as much of the variation from the mean because of the greater noise.

As mentioned previously, one purpose for finding a regression equation is to estimate the values of β_0 and β_1 in the relationship $y = \beta_0 + \beta_1 x + \varepsilon$. The R^2 value is not a measure of the accuracy of these estimates. It is simply a measure of how well the regression line fits the observed data.

Example 3.3.3 Applying R^2 to Linearizable Models
We calculate the R^2 value for a linearizable model by calculating the R^2 value for the straight line fit to the *transformed* data.

1. In the worksheet **Power**, display the R^2 value for the linear trendline on the graph of $\ln L$ versus $\ln P$ by right-clicking on the trendline, and selecting **Format Trendline ...** $\rightarrow$ **Options** $\rightarrow$ **Display R-squared value on chart**. Note this value is 0.8152, indicating a good-fitting model.

2. Also in the worksheet **Power**, right-click on the power trendline on the graph of V versus P. Select **Format Trendline ...** $\rightarrow$ **Options** $\rightarrow$ **Display R-squared value on chart**. This is the same R^2 value.

Repeat this process for the logarithmic and exponential models derived in Examples 3.2.1 and 3.2.3, respectively. We can now compare how well these models fit the data by comparing their R^2 values:

Logarithmic	Power	Exponential
0.8255	0.8152	0.6864

We see that the power and logarithmic models fit the data very well, with the logarithmic model being slightly better. The exponential model does not fit as well. Thus, based strictly on the R^2 values, we would conclude that the logarithmic model fits the data the best. This agrees with our earlier conclusion.

We warn against blindly using R^2 values to choose a "best" model. These values should be used as only one factor when choosing a "best" model. Other factors that should be considered are the nature of the behavior being analyzed and the simplicity of the model.

For instance, population growth is often exponential. So if we fit curves to some data of a population, we may want to favor an exponential model over other types even if it has a lower R^2 value.

In "Modeling the U.S. Population," Sheldon Gordon (1999) makes the point that "The best choice (of a model) depends on the set of data being analyzed and requires an exercise in judgement, not just computation."

Exercises

3.3.1 Table 3.2 gives the diameter of the trunk at chest height and volume of the wood in several pine trees. Use the trendline function in Excel to model volume in terms of diameter with several different linearizable models and select the "best" one.

Diameter	32	29	24	45	20	30	26	40	24	18
Volume	185	109	95	300	30	125	55	246	60	15

Table 3.2

3.3.2 Table 3.3 gives the number of manatee deaths in Florida believed to be caused by watercraft for the years 1983–2006 where Year 1 corresponds to 1983 (Florida Fish and Wildlife Conservation Commission, 2009). Use different linearizable models to model deaths in terms of year and select the "best" one. Is comparing R^2 values sufficient when selecting a "best" model?

3.3 Coefficient of Determination 63

Year	1	2	3	4	5	6	7	8	9	10	11	12
Deaths	15	24	44	44	39	43	50	47	53	38	35	49
Year	13	14	15	16	17	18	19	20	21	22	23	24
					82	78	81	95	73	69	79	92

Table 3.3

r regression line fits a set of data is the *standard error*

$$\sqrt{\frac{\sum (y_i - \hat{y})^2}{n-2}}.$$

$$= \sqrt{\frac{SS_{Res}}{n-2}}.$$

antity is also an estimate of the standard deviation of $x + \varepsilon$.

or the linear model fit to the data with some noise.
n of the "noise"? Try different values of the standard

regression equation fit to a set of data $\{(x_i, y_i) : i = $
d off this equation.

these two properties:

$$= 0 \quad \text{and} \quad \sum_{i=1}^{n} x_i(y_i - \hat{y}_i) = 0.$$

Predicted," calculate $(y_i - \hat{y}_i)$ for each value of y, and
aɪ̣ụᴏ, aᴜᴅ a column titled "x(y – Predicted)," calculate $x_i(y_i - \hat{y}_i)$
for each data pair, and sum all of these values. Note that these sums equal 0.

2. Also modify the worksheet **R2** to illustrate this property:

$$\sum_{i=1}^{n} (y_i - \bar{y})^2 = \sum_{i=1}^{n} (\hat{y}_i - \bar{y})^2 + \sum_{i=1}^{n} (y_i - \hat{y}_i)^2.$$

In other words, show that

$$SS_{Tot} = SS_{Reg} + SS_{Res}.$$

Suggestions: Add a column to calculate SS_{Reg} similar to the columns used to calculate SS_{Res} and SS_{Tot}. Calculate $SS_{Reg} + SS_{Res}$ and note that the sum equals SS_{Tot}.

3. Illustrate that none of these three properties holds if we use a slope other than $\hat{\beta}_1$ or a y-intercept other than $\hat{\beta}_0$ to calculate $\hat{y}_i$. In other words, show that these properties do not hold if we use a slope other than that given by the Excel formula **SLOPE** or a y-intercept other than that given by **INTERCEPT** to calculate the predicted value of y.

4. Use the properties in part (1) to prove the property in part (2). **Hint:** Start with

$$\sum_{i=1}^{n} (y_i - \bar{y})^2 = \sum_{i=1}^{n} (y_i - \bar{y} + \hat{y}_i - \hat{y}_i)^2 = \sum_{i=1}^{n} ((\hat{y}_i - \bar{y}) + (y_i - \hat{y}_i))^2.$$

Then expand the right-hand side and rewrite so that you get the terms $\sum_{i=1}^{n}(\hat{y}_i - \bar{y})^2$, $\sum_{i=1}^{n}(y_i - \hat{y})^2$, $\sum_{i=1}^{n}(y_i - \hat{y}_i)$, and $\sum_{i=1}^{n} x_i (y_i - \hat{y}_i)$.

3.4   Polynomials

Polynomial models are often convenient to use because they are easy to differentiate and integrate. Theorem 3.4.1 is a well-known theorem from algebra about fitting a polynomial model to data.

Theorem 3.4.1 *Given a set of data, $\{(x_i, y_i) : i = 1, \dots, n\}$, where $x_i \neq x_j$ for all $i \neq j$, there exists a unique polynomial $p(x)$ of degree at most $n - 1$ such that*

$$p(x_i) = y_i \text{ for all } i = 1, \dots, n.$$

Graphically, this theorem means that the graph of $y = p(x)$ goes through each data point (i.e., it is a perfect-fitting model). This sounds like a utopia, but is it really?

First, let's take a linear algebra approach to fitting a polynomial to a set of data. Consider the set of data $\{(1, 2), (2, 4), (3, 5)\}$. We will fit a second-degree polynomial of the form $p(x) = ax^2 + bx + c$ to it using matrices. That is, we want values of a, b, and c such that

$$a(1^2) + b(1) + c = 2$$
$$a(2^2) + b(2) + c = 4$$
$$a(3^2) + b(3) + c = 5.$$

This set of equations can be written in matrix form as

$$\begin{bmatrix} 1^2 & 1 & 1 \\ 2^2 & 2 & 1 \\ 3^2 & 3 & 1 \end{bmatrix} \begin{bmatrix} a \\ b \\ c \end{bmatrix} = \begin{bmatrix} 2 \\ 4 \\ 5 \end{bmatrix}.$$

This matrix equation has the generic form $A\vec{x} = \vec{b}$ where $\vec{x} = (a, b, c)$ is the vector of unknowns. Note that the columns of A form a linearly independent set, so A is invertible. Thus there is a unique solution to this matrix equation: $\vec{x} = A^{-1}\vec{b}$.

Example 3.4.1   Solving a Matrix Equation

To find $\vec{x} = A^{-1}\vec{b}$, we first need to calculate A^{-1} and then multiply it by $\vec{b}$. For small matrices, this can be easily done in Excel. (We must note that solving a matrix equation such as this using inverses is computationally expensive. Row-reduction is a more computationally efficient method. Excel will not automatically row-reduce matrices, but it will calculate inverses. For small matrices such as this, computational efficiency is not a real issue.)

1. Rename a blank worksheet "**Poly 3 points**" and format it as in Figure 3.28.

	A	B	C	D	E	F	G	H
1	**Data**				**A**			**b**
2	**x**	**y**		=A3^2	=A3	1		=B3
3	1	2		=A4^2	=A4	1		=B4
4	2	4		=A5^2	=A5	1		=B5
5	3	5						
6					**A⁻¹**			**A⁻¹b**

Figure 3.28

2. To calculate A^{-1}, highlight the range **D7:F9**. Type **=MINVERSE(D2:F4)**, and press the combination of keys **Ctrl-Shift-Enter** (this combination tells Excel to compute an array formula). The result should look as in Figure 3.29.

	D	E	F
6		**A⁻¹**	
7	0.5	-1	0.5
8	-2.5	4	-1.5
9	3	-3	1

Figure 3.29

3. To calculate $A^{-1}\vec{b}$, highlight the range **H7:H9**, type **=MMULT(D7:F9,H2:H4)**, and press **Ctrl-Shift-Enter**. The results are shown in Figure 3.30. This means our polynomial model is $y = -0.5x^2 + 3.5x - 1$.

	H
6	**A⁻¹b**
7	-0.5
8	3.5
9	-1

Figure 3.30

4. Excel will automatically calculate this polynomial for us using an algorithm equivalent to the one described above. Create a scatter plot of the data and right-click on one of the data points, select **Add Trendline**, and add a polynomial curve of degree 2. Under the **Options** tab, select **Display equation on chart**. The results are shown in Figure 3.31. Note that the polynomial it gives is exactly the same as what we calculated and that its graph goes through all three data points.

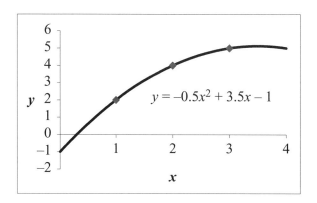

Figure 3.31

Next, we add the point $(4, 2)$ to the data set and fit a second-degree polynomial to these four data points. Ideally, we want to find a, b, and c such that

$$a\left(1^2\right) + b\left(1\right) + c = 2$$
$$a\left(2^2\right) + b\left(2\right) + c = 4$$
$$a\left(3^2\right) + b\left(3\right) + c = 5$$
$$a\left(4^2\right) + b\left(4\right) + c = 2.$$

This set of equations can be written in matrix form as

$$
\begin{bmatrix}
1^2 & 1 & 1 \\
2^2 & 2 & 1 \\
3^2 & 3 & 1 \\
4^2 & 4 & 1
\end{bmatrix}
\begin{bmatrix}
a \\
b \\
c
\end{bmatrix}
=
\begin{bmatrix}
2 \\
4 \\
5 \\
2
\end{bmatrix},
\tag{3.2}
$$

which, as before, has the generic form $A\vec{x} = \vec{b}$. However, note that A has more rows than columns, so it is not invertible. Further analysis reveals that this equation does not even have a solution, so there is no second-degree polynomial that fits these four data points perfectly. We will have to settle for a "best-fit" polynomial model. As in Chapter 2 when we fit a linear model to a set of data, we will use a least-squares criterion to find our model. That is, we want a polynomial $p(x)$ that minimizes the number

$$
\sum_{i=1}^{n} (y_i - p(x_i))^2,
$$

where the sum is taken over all data points. The resulting model is called a *least-squares polynomial model*. To find this model we find a *least-squares solution* to the matrix equation (3.2).

Definition 3.4.1 Let A be an $m \times n$ matrix and $\vec{b} \in R^m$. A *least-squares solution* of $A\vec{x} = \vec{b}$ is a vector $\hat{x} \in R^n$ such that

$$
\left\| \vec{b} - A\hat{x} \right\| \leq \left\| \vec{b} - A\vec{x} \right\| \text{ for all } \vec{x} \in R^n.
$$

The idea behind this definition is that $\hat{x} \in R^n$ gets $A\vec{x}$ as close to $\vec{b}$ as possible. A theorem from linear algebra tells us how to find $\hat{x}$.

Theorem 3.4.2 *Every least-squares solution of $A\vec{x} = \vec{b}$ must satisfy the "normal" equation*

$$
A^T A\vec{x} = A^T \vec{b}.
$$

If the columns of A are linearly independent, then $A^T A$ is invertible and there is a unique least-squares solution $\hat{x}$ given by

$$
\hat{x} = \left(A^T A \right)^{-1} A^T \vec{b}.
\tag{3.3}
$$

Note that Theorem 3.4.2 does not say that a general matrix equation $A\vec{x} = \vec{b}$ has a unique least-squares solution. In general, there may be many least-squares solutions. However, if the columns of A are linearly independent, then the solution is unique.

When fitting curves to data, as in this example, the columns of A are usually linearly independent, so we can use equation (3.3) to calculate $\hat{x}$. This technique can also be used to fit other types of curves to data, as we will see later.

Example 3.4.2   Calculating a Least-Squares Solution

To calculate $\hat{x}$ using equation (3.3) in Excel, we will first calculate A^T, then $A^T A$, then $\left(A^T A\right)^{-1}$, then $\left(A^T A\right)^{-1} A^T$, and finally $\left(A^T A\right)^{-1} A^T \vec{b}$.

1. Rename a blank worksheet "**Poly 4 points**" and format it as in Figure 3.32.

	A	B	C	D	E	F	G	H
1	**Data**				**A**			**b**
2	**x**	**y**		=A3^2	=A3	1		=B3
3	1	2		=A4^2	=A4	1		=B4
4	2	4		=A5^2	=A5	1		=B5
5	3	5		=A6^2	=A6	1		=B6
6	4	2						

Figure 3.32

2. Format the spreadsheet as in Figure 3.33 to hold the various matrix calculations needed.

	A	B	C	D	E	F	G	H
8		**AT**					**ATA**	
9								
10								
11								
12								
13		**(ATA)-1**					**(ATA)-1AT**	
14								
15								
16								
17								
18		**(ATA)-1ATb**						

Figure 3.33

3. Highlight range **A9:D11**, type =**TRANSPOSE(D2:F5)**, and press **Ctrl-Shift-Enter**.

4. Highlight range **F9:H11**, type =**MMULT(A9:D11,D2:F5)**, and press **Ctrl-Shift-Enter**.

5. Highlight range **A14:C16**, type =**MINVERSE(F9:H11)**, and press **Ctrl-Shift-Enter**.

6. Highlight range **E14:H16**, type =**MMULT(A14:C16,A9:D11)**, and press **Ctrl-Shift-Enter**.

7. Highlight range **B19:B21**, type =**MMULT(E14:H16,H2:H5)**, and press **Ctrl-Shift-Enter**. The results are shown in Figure 3.34. This means our polynomial model is $y = -1.25x^2 + 6.35x - 3.25$. Such a model is called a *least-squares* second-degree polynomial model.

	B
18	$(A^TA)^{-1}A^Tb$
19	-1.25
20	6.35
21	-3.25

Figure 3.34

8. Create a graph of the data points, add a second-degree polynomial trendline, and display the equation on the chart as in Figure 3.35. Note that this model is exactly the same as what we calculated.

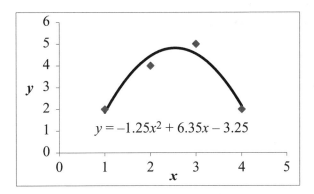

Figure 3.35

Example 3.4.3   Calculating R^2 Value

We calculated R^2 values for linearizable models by calculating the R^2 value for the straight line fit to the transformed data. Polynomial models are not linearizable, so to calculate R^2 values for these types of models, we must use the definition. The definition is

$$R^2 = \frac{SS_{Tot} - SS_{Res}}{SS_{Tot}}$$

where $SS_{Tot} = \sum (y_i - \bar{y})^2$ and $SS_{Res} = \sum (y_i - \hat{y}_i)^2$. Remember that $\bar{y}$ is the mean of all the y-values in the data set and that $\hat{y}_i$ is the *predicted* value of y_i based on the model. Here we will use our polynomial model $y = -1.25x^2 + 6.35x - 3.25$ to calculate $\hat{y}_i$.

1. Rename a blank worksheet "**Poly R2**" and format it as in Figure 3.36. The R^2 value is 0.9333, indicating a very good fit.

	A	B	C	D	E
1	**x**	**y**	**Predicted**	**SS$_{Tot}$**	**SS$_{Res}$**
2	1	2	=-1.25*A2^2+6.35*A2-3.25	=(B2-$B$6)^2	=(B2-C2)^2
3	2	4	=-1.25*A3^2+6.35*A3-3.25	=(B3-$B$6)^2	=(B3-C3)^2
4	3	5	=-1.25*A4^2+6.35*A4-3.25	=(B4-$B$6)^2	=(B4-C4)^2
5	4	2	=-1.25*A5^2+6.35*A5-3.25	=(B5-$B$6)^2	=(B5-C5)^2
6	**Mean =**	=AVERAGE(B2:B5)	**Totals =**	=SUM(D2:D5)	=SUM(E2:E5)
7				**R² =**	=(D6-E6)/D6

Figure 3.36

2. Excel will automatically calculate this R^2 value. On the graph of y versus x in the worksheet **Poly 4 points**, add the R^2 value to the polynomial trendline by right-clicking on the trendline and selecting **Format Trendline** ... → **Options** → **Display R-squared value on chart**. This value is equal to what we calculated.

Example 3.4.4   Selecting a "Best" Polynomial Model

Consider the data in Table 3.4 which gives the area A (in thousands of square miles) and the total length of railroad track R (in thousands of miles) of seven different countries (*The World Almanac and Book of Facts*, 2007). We want to use a polynomial to model R in terms of A.

According to Theorem 3.4.1, there is a unique polynomial of degree at most 6 that fits this data perfectly. We can easily graph R versus A and fit a sixth-degree polynomial trendline. The result is shown in Figure 3.37.

	Luxembourg	Ireland	Azerbaijan	S. Korea	Greece	Finland	Japan
A	0.998	27.135	33.436	38.023	50.942	130.559	292.26
R	0.170	2.058	1.834	2.157	1.598	3.635	4.092

Table 3.4

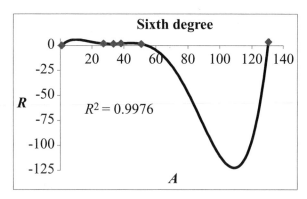

Figure 3.37

Notice that the R^2 value is 1 (within rounding error), so the model fits the data perfectly. However, if we were to use this model to make predictions, it would predict that a country with 100,000 square miles ($A = 100$) would have approximately $-125,000$ miles of railroad track ($R = -125$). This is totally unreasonable, so this is a terrible model even though it fits the data perfectly. The large oscillation seen in the graph of this model is typical for a "high-order" polynomial model such as this.

To find a better model, we can fit polynomials of order 1 through 5 to the data (note that a first-order polynomial model is a linear model), and keep track of their R^2 values. The graphs of these models are shown in Figure 3.38.

To choose the "best" model, we need to examine more than just the R^2 values. We also need to consider how well they will make predictions and their simplicity. The fourth- and fifth-degree models have the highest R^2 values, but they both have oscillations that seem unreasonable in the context of the problem, so they are not the best options. The first-degree model does not capture the trend of the data very well, so it is not a good option either. The second- and third-degree models have very similar R^2 values and their graphs look very similar. So we will choose the simpler of the two options, the second-degree model, as the "best." However, one could make a case that the third-degree model is the "best."

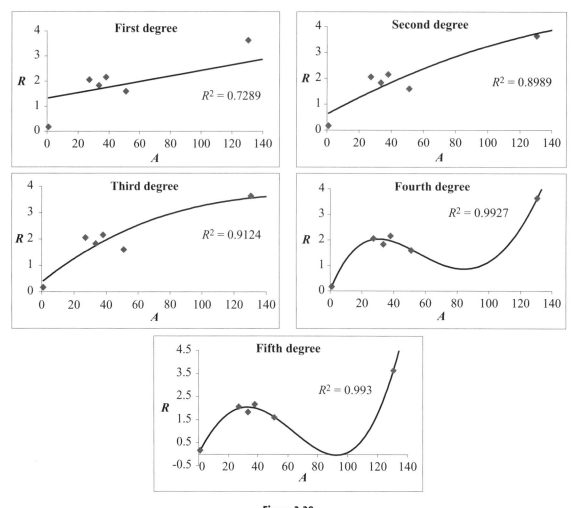

Figure 3.38

Example 3.4.5   Deer Population

Table 3.5 gives the number of deer in a hypothetical forest for various years between 1941 and 1982. Our goal is to model the population in terms of the year.

Year	1941	1947	1951	1957	1962	1965	1971	1977	1982
Population	12,500	28,500	7000	20,000	6500	12,000	4000	11,000	3500

Table 3.5

Examining the graph of the data in Figure 3.39 we see that the population fluctuates from a high to a low and back to a high in about a 10-year cycle.

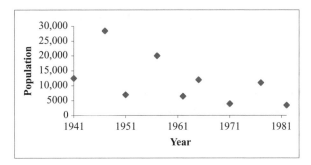

Figure 3.39

The periodic nature of the data suggests we use sine or cosine functions to model it. Since the period is approximately 10, our model will have terms of $\cos\left(\frac{\pi x}{5}\right)$ and $\sin\left(\frac{\pi x}{5}\right)$ (where x represents the year and y represents the population). Also note that the population shows a downward trend. Therefore, we include an x term in the model with a (likely) negative coefficient. Thus our model is of the form

$$y = a + bx + c\cos\left(\frac{\pi x}{5}\right) + d\sin\left(\frac{\pi x}{5}\right).$$

Ideally we want to satisfy the system of equations

$$a + 1941b + c\cos\left(\frac{1941\pi}{5}\right) + d\sin\left(\frac{1941\pi}{5}\right) = 12{,}500$$

$$a + 1947b + c\cos\left(\frac{1947\pi}{5}\right) + d\sin\left(\frac{1947\pi}{5}\right) = 28{,}500$$

$$\vdots$$

$$a + 1982b + c\cos\left(\frac{1982\pi}{5}\right) + d\sin\left(\frac{1982\pi}{5}\right) = 3500,$$

which has the matrix form

$$\begin{bmatrix} 1 & 1941 & \cos\left(\frac{1941\pi}{5}\right) & \sin\left(\frac{1941\pi}{5}\right) \\ 1 & 1947 & \cos\left(\frac{1947\pi}{5}\right) & \sin\left(\frac{1947\pi}{5}\right) \\ \vdots & \vdots & \vdots & \vdots \\ 1 & 1982 & \cos\left(\frac{1982\pi}{5}\right) & \sin\left(\frac{1982\pi}{5}\right) \end{bmatrix} \begin{bmatrix} a \\ b \\ c \\ d \end{bmatrix} = \begin{bmatrix} 12{,}500 \\ 28{,}500 \\ \vdots \\ 3500 \end{bmatrix}.$$

As before, this matrix equation has the generic form $A\vec{x} = \vec{b}$. Calculating the least-squares solution, $\hat{x} = \left(A^T A\right)^{-1} A^T \vec{b}$, to this system as done in Example 3.4.2, gives the approximate solution $\hat{x} = (628,860, -314, -2892, -5704)$. So the model is

$$y = 628,860 - 314x - 2892\cos\left(\frac{\pi x}{5}\right) - 5704\sin\left(\frac{\pi x}{5}\right).$$

Figure 3.40 shows a graph of the model on top of the data. The model appears to fit the data relatively well, but many refinements could be made. In the exercises you will be asked to make one such refinement.

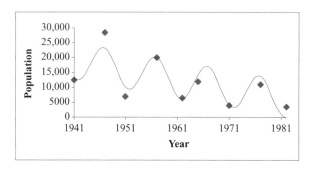

Figure 3.40

General Guidelines for Selecting a "Best" Model

In general when constructing empirical models, one has many different types of models from which to choose. Deciding which one is "best" is not easy and is often very subjective, but here are a few simple guidelines:

1. Consider the R^2 value, but don't rely solely on it.
2. Look for a pattern in the residuals. If there is a pattern, the model should be refined.
3. Consider how good the model is for making predictions between data values. If it "oscillates" or would give unreasonable values, look for a better model.
4. Consider "end" behavior. If the data appears to be "leveling off" at the end, but the model is increasing (or vice versa), consider a different model.
5. Consider the simplicity of the model. In general, the fewer the number of terms, the better.

We also stress that when using an empirical model such as those discussed here to make predictions, the predictions are always point-*estimates* of the true values. These predictions should never be presented as precise certainties. Books on statistics and regression discuss how to use a point-estimate to form a confidence interval for the true value (a range of possible values), but this topic is beyond the scope of this book.

Exercises

3.4.1 Use the least-squares matrix approach to fit a linear model $y = a + mx$ to a randomly generated set of data with four points. Show by examples that this gives the same results as the formulas given in Chapter 2.

3.4.2 Fit different polynomial models to the data in Table 3.6 and select the "best" one. Explain how you determined which one is "best."

x	1.4	2.4	7.1	13.8	34.2	109.3	134
y	2.7	2.27	3.31	3.39	3.81	4.88	4.62

Table 3.6

3.4.3 Calculate the R^2 value for the model fit to deer population data in Example 3.4.5.

3.4.4 Consider the problem of fitting a second-degree polynomial to a set of four data points discussed in Example 3.4.2. Add the requirement that $y(0) = c_0$ where c_0 is some specified value (in other words, we are specifying the y-intercept of the model). Design a spreadsheet to find a least-squares second-degree polynomial model where the user can input four data points and specify the value of c_0.

3.4.5 The amount of a radioactive substance remaining after time $y(t)$ is described by the exponential model $y(t) = Ce^{-kt}$ where C is the initial amount (the amount at time $t = 0$) and k is a constant. Suppose two radioactive substances A and B have constants $k_A = 0.03$ and $k_B = 0.05$, respectively. A mixture of these two substances contains C_A grams of A and C_B grams of B at time $t = 0$, both of which are unknown. The total amount of the mixture at time t is modeled by

$$y(t) = C_A e^{-0.03t} + C_B e^{-0.05t}. \tag{3.4}$$

A researcher measures the total amount of substance at several times and records the data in Table 3.7. Estimate the values of C_A and C_B by fitting a least-squares model of the form (3.4) to the data.

Time	5	6	7	8
Amount	8.8	8.6	8.2	7.9

Table 3.7

3.4.6 Consider a refinement to the deer population model in Example 3.4.5. Note that as time increases, the difference between a high point and the next low point (the amplitude) tends to decrease. Our original model did not take this into account.

1. For the years 1947, 1957, 1965, and 1977 calculate the amplitude by subtracting the population in the next year.

2. Create a graph of amplitude versus year using the data in part (1).

3. Fit an exponential model, $g(x) = me^{kx}$, to the data in part (1).

4. Fit a model of the form

$$y = a + bx + c\,g(x) * \cos\left(\frac{\pi x}{5}\right) + d\,g(x) * \sin\left(\frac{\pi x}{5}\right)$$

to the original data. Does this model seem to fit the data any better than the original one?

5. Calculate the R^2 value for this refined model. How does this compare to the original model?

3.4.7 Table 3.8 contains the temperatures over one day in Seward, Nebraska, starting at midnight.

Hour	0	2	4	6	8	10	12	14	16	18	20	22
Temp	44.5	45.3	52.6	60.4	70.2	75.9	79.8	79.1	72.8	63.5	52.5	44.6

Table 3.8

Our goal is to predict the temperature y at time x. Notice that the temperature is periodic, so we will use a model of the form $y = a \sin(bx + c) + d$. Follow these steps to design a spreadsheet to implement a simple "sine regression" algorithm for fitting a model of this type to the data:

1. Enter the data in a spreadsheet, create a graph of the data, and designate cells to hold the values of a, b, c, and d.

2. The period of the function $y = a \sin(bx + c) + d$ is $\frac{2\pi}{b}$. Estimate the period of the data and use this to estimate the value of b.

3. Initially, let $c = 0$.

4. To find the values of a and d, create a graph of y versus $\sin(bx + c)$ and fit a linear trendline to this transformed data. Display the R^2 value. The slope is the value of a and the y-intercept is the value of d. Use the functions **SLOPE** and **INTERCEPT** to calculate the values of a and d, respectively.

5. Create a scroll bar to vary the value of b between -1 and $+1$ and another scroll bar to vary the value of c between -2 and $+2$.

6. Use the scroll bars to find values of b and c that maximize the R^2 value.

7. Graph the model on top of the original data. How well does the model fit the data?

3.5   Multiple Regression

In previous sections we have discussed predicting the value of one response variable y with one predictor variable x. In this section we discuss using two or more predictor variables $x_1, x_2, \ldots, x_n$. This topic is called *multiple regression*.

Consider the problem of predicting the selling price of a house. The selling price is affected by many factors, including the age of the house, living area, number of bedrooms, and so on. Table 3.9 lists the selling price, living area (in ft^2), acres of land, and the number of bedrooms of 10 homes in a neighborhood.

Selling Price ($)	Area (ft^2)	Acres	Bedrooms
100,000	2205	2.5	3
93,500	2155	0.8	3
95,650	2600	1.1	4
75,025	1900	0.35	3
95,000	1200	2.5	2
80,250	2050	1.8	3
85,250	2250	0.9	4
121,250	2490	1.8	3
94,575	2390	1.6	2
109,000	3100	1.0	4

Table 3.9

Example 3.5.1   Single Predictor Variable

Consider the graphs of Selling Price versus Area, and Selling Price versus Acres, as shown in Figure 3.41 along with the linear regression equation for each.

Notice that the R^2 value for the predictor variable area is higher than the R^2 value for acres. This means that the regression equation for selling price in terms of area will give a

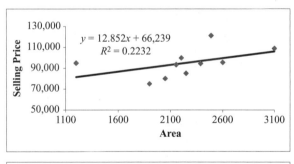

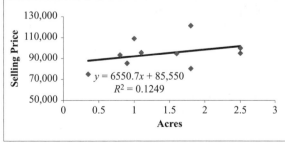

Figure 3.41

better predicted selling price than the equation in terms of acres. We say that out of these two variables, taxes is the "best" single-variable predictor of selling price.

Suppose a house has an area of 2800 ft^2. We could use the regression equation for selling price in terms of area to predict that the house would sell for $12.852(2800) + 66,239 = \$102,224$. This, of course, is only a point-estimate of the price, and not a very good estimate because the R^2 value for the regression equation is only 0.2232.

Example 3.5.2   Multiple Predictor Variables

Considering only one predictor variable is a bit too simple. Many variables affect the selling price, so we should consider more than one predictor variable in our regression equation. Suppose we consider both area and acres.

If we let y = selling price, x_1 = area, and x_2 = acres, we want to fit a model of the form

$$y = a_0 + a_1 x_1 + a_2 x_2$$

to the data where a_0, a_1, and a_2 are constants. As in Section 3.4, ideally we want to satisfy the system of equations

$$a_0 + a_1 2{,}205 + a_2 2.5 = 100{,}000$$

$$\vdots$$

$$a_0 + a_1 3{,}100 + a_2 1 = 109{,}000.$$

We could take a matrix approach to find a least-squares solution to this system as we did in Section 3.4, but Excel will do this automatically for us.

1. Rename a blank worksheet "**Homes**" and format it as in Figure 3.42. Enter the rest of the data from Table 3.9 in columns **A–D**.

	A	B	C	D
1	**Selling Price**	**Area**	**Acres**	**Bedrooms**
2	100000	2205	2.5	3

Figure 3.42

2. Select **Tools → Data Analysis... → Regression** and click **OK**. (**Note:** If **Data Analysis...** is not available, select **Tools → Add-Ins... → Analysis ToolPak**, click **OK**, and try selecting **Data Analysis...** again.)

   (a) Next to **Input Y Range:**, select **$A$1:$A$11**.

(b) Next to **Input X Range:**, select **$B$1:$C$11**.

(c) Check the box next to **Labels**.

(d) Next to **Output Range:**, select **$A$13**.

(e) Click **OK**.

Excel generates many outputs. We focus on two subsets. (For a more complete coverage of the outputs see, for example, Sincich, Levine, and Stephan, 2002.) The first subset of outputs shown in Figure 3.43 gives the coefficients in our model (i.e., the values of a_0, a_1, and a_2). We see that our model is

$$y = 36{,}669.6 + 18.9x_1 + 11{,}239.2x_2.$$

Such a model is called a *multiple-regression equation*.

	A	B
28		Coefficients
29	Intercept	36669.57782
30	Area	18.86847037
31	Acres	11239.20513

Figure 3.43

The second subset of outputs is shown in Figure 3.44. Here we see the R^2 value that is calculated by equation (3.1). The other important output is the adjusted R^2 value. The adjusted R^2 value is calculated using the formula

$$\text{adjusted } R^2 = 1 - \left[\frac{n-1}{n - (k+1)} \right] (1 - R^2),$$

where n is the number of data points and k is the number of predictor variables ($n = 10$ and $k = 2$ in this case).

	A	B
15	Regression Statistics	
16	Multiple R	0.736297484
17	R Square	0.542133985
18	Adjusted R Square	0.411315124
19	Standard Error	10308.18525
20	Observations	10

Figure 3.44

The adjusted R^2 value takes into account the number of data points (the more data points, the higher the adjusted R^2 value) and the number of predictor variables (the more predictor variables, the lower the adjusted R^2 value). We want as simple a model as possible, so the fewer the variables, the better. We will compare different sets of predictor variables using adjusted R^2 values.

3. Now let's consider the combination of predictor variables area and bedrooms. Repeat Step 2, except select **$C$1:$D$11** for the **Input X Range:**. The R^2 and adjusted R^2 values for this combination are 0.327 and 0.135, respectively, which are slightly lower than the previous combination, indicating that this combination does not give a better model for predicting the selling price.

The results for all the different combinations of predictor variables are shown in Table 3.10. Note that including the variable bedrooms results in low R^2 values. This indicates that the number of bedrooms is not a good predictor of the selling price. Also note that the R^2 value for the set of all three predictor variables is the highest, but the adjusted R^2 value is lower than that for area and acres. This indicates that the set of three variables gives better predictions (a higher R^2 value), but the additional variable makes it more complicated, so it is less desirable as a model (a lower adjusted R^2 value).

Predictor Variables	R^2	Adjusted R^2
Area, Acres	0.542	0.411
Area, Bedrooms	0.327	0.135
Acres, Bedrooms	0.209	−0.017
Area, Acres, Bedrooms	0.552	0.328

Table 3.10

If we simply compare the adjusted R^2 values, we conclude that the "best" combination of predictor variables is area and acres. To refine our model we might want to collect data on other variables that might affect the selling price, such as age, total number of rooms, and so on. With more variables, there are many different combinations of predictor variables we could consider. The process of determining which set of variables is "best" is very complicated. We have presented a very simple strategy here.

Example 3.5.3   Polynomial Models

Let's return to the problem of fitting a second-degree polynomial model of the form $y = a + bx + cx^2$ to the set of four data points $\{(1, 2), (2, 4), (3, 5), (4, 2)\}$ as considered in Example 3.4.2. We could think of this as predicting the values of y by using the "predictor" variables x and x^2, so it can be treated as a multiple regression problem.

1. Rename a blank worksheet "**Polynomial**" and format it as in Figure 3.45. Enter the rest of the data in columns **A** and **B** and copy the formula in **C3** down to row 6.

	A	B	C
1	**Data**		
2	**y**	**x**	**x²**
3	2	1	=B3^2

Figure 3.45

2. Repeat Step 2 from Example 3.5.2. Select **$A$2:$A$6** as the **Input Y Range:** and **$B$2:$C$6** as the **Input X Range:**. Check the box next to **Labels** and select **$A$8** as the **Output Range:**. The coefficients are shown in Figure 3.46.

	A	B
23		*Coefficients*
24	Intercept	-3.25
25	x	6.35
26	x2	-1.25

Figure 3.46

These results give us the model $y = -3.25 + 6.35x - 1.25x^2$, which is exactly the same as in Example 3.4.2. Also note that the R^2 value is 0.9333, exactly the same as that calculated in Example 3.4.3.

Example 3.5.4   Deer Population Again

We can use a multiple regression approach to fit a model of the form $y = a + bx + c\cos\left(\frac{\pi x}{5}\right) + d\sin\left(\frac{\pi x}{5}\right)$ to the deer population data given in Table 3.5. In this case we have three "predictor" variables: x, $\cos\left(\frac{\pi x}{5}\right)$, and $\sin\left(\frac{\pi x}{5}\right)$.

1. Rename a blank worksheet "**Deer Population**" and format it as in Figure 3.47. Enter the rest of the data from Table 3.5 in columns **A** and **B** and copy the range **C2:D2** down to row 10.

	A	B	C	D
1	**Population**	**Year**	**cos**	**sin**
2	12500	1941	=COS(PI()*B2/5)	=SIN(PI()*B2/5)

Figure 3.47

2. Repeat Step 2 from Example 3.5.2. Select **$A$1:$A$10** as the **Input Y Range:** and **$B$1:$D$10** as the **Input X Range:**. Check the box next to **Labels** and select **$A$12** as the **Output Range:**. The coefficients are shown in Figure 3.48.

	A	B
27		Coefficients
28	Intercept	628860.1703
29	Year	-314.2170872
30	cos	-2892.334521
31	sin	-5704.2569

Figure 3.48

These coefficients give us the approximate model

$$y = 628860 - 314x - 2892\cos\left(\frac{\pi x}{5}\right) - 5704\sin\left(\frac{\pi x}{5}\right),$$

which is exactly the same as in Example 3.4.5. Also note that the R^2 value is 0.8771, which should be (approximately) the same as that calculated in Exercise 3.4.3.

Exercises

3.5.1 In an attempt to predict the final grade of students in an Introduction to Statistics class, the professor gives each student a 20-point pretest at the beginning of the year. Table 3.11 gives the final grade (4 = A, 3 = B, etc.), pretest score, ACT score, and year (1 = freshman, 2 = sophomore, etc.) of 10 students.

Grade	Pretest	ACT	Year
2	9	25	1
2	8	20	2
1	18	18	2
1	10	17	4
1	6	20	3
1	8	22	3
3	16	30	1
2	11	28	4
3	15	27	4
4	19	31	3

Table 3.11

1. Determine which variable—pretest score, ACT score, or year—is the best single-variable predictor of the final grade based on the adjusted R^2 values. Does the year of the student appear to be related to the grade at all?

2. Consider all four different combinations of two or three predictor variables. Use the adjusted R^2 values to determine which combination is "best" at predicting the final grade. Based on your results, does it seem worthwhile to give the pretest as a way of predicting the final grade?

3. Use the multiple regression equation that predicts grade in terms of pretest and ACT scores to predict the grade of a student who has a pretest score of 18 and an ACT score of 28.

3.5.2 Fit a seventh-degree polynomial to the eight data points shown in Table 3.12. Create a graph of the resulting polynomial on top of the data points.

x	0.50	0.55	0.60	0.65	0.70	0.75	0.80	0.85
y	0.90	9.00	2.10	7.80	13.50	9.30	0.40	6.20

Table 3.12

3.6   Spline Models

Consider the data in Table 3.13 that gives the liters of milk given by a dairy cow on each of several different days after she begins producing. A graph of this data is shown in Figure 3.49. Our goal is to model liters in terms of day so that we can predict how much milk was given on the days not listed in the table.

Day	5	13	25	42	50	62	75	90	100	120
Liters	10	19	30	25	22	35	50	55	40	35

Table 3.13

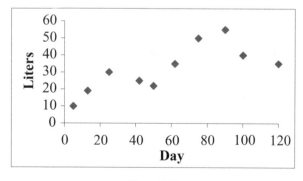

Figure 3.49

The graph of the data certainly does not resemble an exponential, logarithmic, or power curve, so a linearizable model does not seem appropriate. Low-order polynomials do not capture the trend of the data, and higher-order polynomials produce oscillations that do not seem appropriate. We instead consider *spline* models where we simply "connect the dots" with either straight lines, forming a *linear spline* model, or with cubic polynomials, forming a *cubic spline* model.

Example 3.6.1   Linear Spline Model

The graph of a linear spline model is easy to form.

1. Rename a blank worksheet "**Linear**" and format it as in Figure 3.50. Enter the rest of the data from Table 3.13 in columns **A** and **B**.

	A	B
1	Day	Liters
2	5	10

Figure 3.50

2. Create a graph similar to Figure 3.51. The straight lines form the graph of the linear spline model.

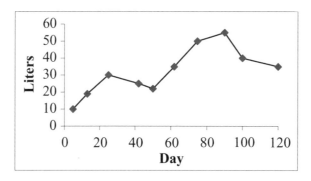

Figure 3.51

3. To use this model to make predictions, we need to know the slope and y-intercept of each line segment. We know two points on each line segment—call them (x_n, y_n) and (x_{n+1}, y_{n+1}) where $x_n < x_{n+1}$—so we can easily find the slope using the formula

$$\text{slope} = m = \frac{y_{n+1} - y_n}{x_{n+1} - x_n}.$$

Once we know the slope m we can find the y-intercept b using the slope-intercept form of a straight line:

$$y = mx + b \quad \Rightarrow \quad b = y - mx.$$

To implement these formulas, format the spreadsheet as in Figure 3.52. Copy row 2 down to row 10.

	C	D
1	**Slope**	**Intercept**
2	=(B3–B2)/(A3–A2)	=B2–C2*A2

Figure 3.52

Looking at the column **Slope**, we see that the amount of milk produced is increasing most rapidly, on average, between days 62 and 75 and decreasing most rapidly between days 90 and 100.

4. Now that we know the slopes and y-intercepts of each piece of the spline we can easily calculate a predicted value of liters given a value of days by

$$\text{liters} = m(\text{days}) + b$$

where m and b are the slope and y-intercept, respectively, of the appropriate piece of the spline. To do this, add the formulas in Figure 3.53. (**Note:** The function **VLOOKUP** in Figure 3.53 will look down the left column of the range **A2:D10** and find the largest value less than or equal to the value in cell **F2**. It will then return the value in the third or fourth column of that range, the slope or the y-intercept, respectively. For this to work properly it is necessary that the x-values are in ascending order.)

	F	G
1	Day	Liters
2	12	=VLOOKUP(F2,A2:D10,3)*F2+VLOOKUP(F2,A2:D10,4)

Figure 3.53

Example 3.6.2   Cubic Spline Model

Linear spline models are easy to calculate, but they do not form smooth curves. In fact, the curve is not differentiable at any one of the data points. This type of model predicts sharp changes at each data point that do not seem realistic. To solve this problem, we connect the dots with cubic polynomials instead of straight lines and put conditions on the derivatives of each segment.

To illustrate this process, we consider a set of three data points $\{(x_1, y_1), (x_2, y_2), (x_3, y_3)\}$ where $x_1 < x_2 < x_3$. Each x-value is called a *knot*. We connect them using two cubic polynomials:

$$p_1(x) = a_1 + b_1 x + c_1 x^2 + d_1 x^3 \quad \text{for } x_1 \le x < x_2$$
$$p_2(x) = a_2 + b_2 x + c_2 x^2 + d_2 x^3 \quad \text{for } x_2 \le x \le x_3.$$

We now need to find the values of the eight parameters a_1, b_1, c_1, d_1, a_2, b_2, c_2, and d_2. For the model to form a smooth curve that goes through each data point, we need to satisfy the following three conditions:

1. Each polynomial must pass through the two data points at the ends of the interval over which it is defined.

2. The first derivatives of two polynomials that meet must be equal at the point at which they meet.

3. The second derivatives of two polynomials that meet must be equal at the point at which they meet.

The first condition gives us the following four equations:

$$p_1(x_1) = a_1 + b_1 x_1 + c_1 x_1^2 + d_1 x_1^3 = y_1 \tag{3.5}$$
$$p_1(x_2) = a_1 + b_1 x_2 + c_1 x_2^2 + d_1 x_2^3 = y_2 \tag{3.6}$$
$$p_2(x_2) = a_2 + b_2 x_2 + c_2 x_2^2 + d_2 x_2^3 = y_2 \tag{3.7}$$
$$p_2(x_3) = a_2 + b_2 x_3 + c_2 x_3^2 + d_2 x_3^3 = y_3. \tag{3.8}$$

The first derivatives of the polynomials are

$$p_1'(x) = b_1 + 2c_1 x + 3d_1 x^2, \quad p_2'(x) = b_2 + 2c_2 x + 3d_2 x^2.$$

The second condition gives us the equation

$$p_1'(x_2) = b_1 + 2c_1 x_2 + 3d_1 x_2^2 = b_2 + 2c_2 x_2 + 3d_2 x_2^2 = p_2'(x_2). \tag{3.9}$$

The second derivatives of the polynomials are

$$p_1''(x) = 2c_1 + 6d_1 x, \quad p_2''(x) = 2c_2 + 6d_2 x.$$

The third condition gives us the equation

$$p_1''(x_2) = 2c_1 + 6d_1 x_2 = 2c_2 + 6d_2 x_2 = p_2''(x_2). \tag{3.10}$$

Equations (3.5)–(3.10) give us six equations for the eight unknowns. To uniquely determine the values of these unknowns, we need two more equations. To this end, we specify the values of the second derivatives at the end points of the data set (at x_1 and x_3).

If we want these second derivatives to be some known values, say m_1 and m_2, we would add the equations

$$p_1''(x_1) = 2c_1 + 6d_1x_1 = m_1$$
$$p_2''(x_3) = 2c_2 + 6d_2x_3 = m_2.$$

The resulting model is called a *clamped spline*. If, however, we do not know the values of the derivatives, we simply set them equal to 0 and have the equations

$$p_1''(x_1) = 2c_1 + 6d_1x_1 = 0 \tag{3.11}$$
$$p_2''(x_3) = 2c_2 + 6d_2x_3 = 0. \tag{3.12}$$

This is the approach we will take. The resulting model is called a *natural spline*. Rewriting equations (3.5)–(3.12) so that the constants are on the right-hand side, we get the system

$$
\begin{array}{llll}
a_1 + x_1b_1 + x_1^2c_1 + x_1^3d_1 & & & = y_1 \\
a_1 + x_2b_1 + x_2^2c_1 + x_2^3d_1 & & & = y_2 \\
& a_2 + x_2b_2 + x_2^2c_2 + x_2^3d_2 & & = y_2 \\
& a_2 + x_3b_2 + x_3^2c_2 + x_3^3d_2 & & = y_3 \\
b_1 + 2x_2c_1 + 3x_2^2d_1 & - & b_2 - 2x_2c_2 - 3x_2^2d_2 & = 0 \\
2c_1 + 6x_2d_1 & - & 2c_2 - 6x_2d_2 & = 0 \\
2c_1 + 6x_1d_1 & & & = 0 \\
& & 2c_2 + 6x_3d_2 & = 0.
\end{array}
$$

Rewriting this system in matrix form yields

$$
\begin{bmatrix}
1 & x_1 & x_1^2 & x_1^3 & 0 & 0 & 0 & 0 \\
1 & x_2 & x_2^2 & x_2^3 & 0 & 0 & 0 & 0 \\
0 & 0 & 0 & 0 & 1 & x_2 & x_2^2 & x_2^3 \\
0 & 0 & 0 & 0 & 1 & x_3 & x_3^2 & x_3^3 \\
0 & 1 & 2x_2 & 3x_2^2 & 0 & -1 & -2x_2 & -3x_2^2 \\
0 & 0 & 2 & 6x_2 & 0 & 0 & -2 & -6x_2 \\
0 & 0 & 2 & 6x_1 & 0 & 0 & 0 & 0 \\
0 & 0 & 0 & 0 & 0 & 0 & 2 & 6x_3
\end{bmatrix}
\begin{bmatrix}
y_1 \\
y_2 \\
y_2 \\
y_3 \\
0 \\
0 \\
0 \\
0
\end{bmatrix}
=
,
\tag{3.13}
$$

which has the general form $A\vec{x} = \vec{b}$ where $\vec{x} = (a_1, b_1, c_1, d_1, a_2, b_2, c_2, d_2)$ is the vector of unknowns. We can solve this system using inverses, $\vec{x} = A^{-1}\vec{b}$. (Note that matrix A has a distinct structure and a lot of 0s. These facts can be exploited to find more computationally efficient ways to solve this system.)

To fit a cubic spline to the three data points $\{(75, 50), (95, 55), (100, 40)\}$ from Table 3.13, follow these steps:

1. Rename a blank worksheet "**Cubic**" and format it as in Figure 3.54.

	A	B	C	D	E	F	G	H	I	J	K	L	M
1	x	y					A						b
2	75	50		1	=A2	=A2^2	=A2^3	0	0	0	0		=B2
3	90	55		1	=A3	=A3^2	=A3^3	0	0	0	0		=B3
4	100	40		0	0	0	0	1	=A3	=A3^2	=A3^3		=B3
5				0	0	0	0	1	=A4	=A4^2	=A4^3		=B4
6				0	1	=2*A3	=3*A3^2	0	=-E6	=-F6	=-G6		0
7				0	0	2	=6*A3	0	0	=-F7	=-G7		0
8				0	0	2	=6*A2	0	0	0	0		0
9				0	0	0	0	0	0	2	=6*A4		0
10							A^{-1}						$A^{-1}b$

Figure 3.54

2. To calculate A^{-1}, highlight the range **D11:K18**. Type **=MINVERSE(D2:K9)**, and press the combination of the keys **Ctrl-Shift-Enter**. To calculate $A^{-1}\vec{b}$, highlight the range **M11:M18**. Type **=MMULT(D11:K18,M2:M9)**, and press the combination of the keys **Ctrl-Shift-Enter**. The results give the model

$$p_1(x) = 1015 - 40.37x + 0.55x^2 - 0.0024x^3 \quad \text{for } 75 \le x < 90$$
$$p_2(x) = -3440 + 108.13x - 1.1x^2 + 0.0037x^3 \quad \text{for } 90 \le x \le 100.$$

3. To graph the resulting model, add the formulas in Figure 3.55. Copy row 9 down to row 58. Use these results, along with the original data, to form a graph similar to Figure 3.56.

	A	B
6		Spline
7	x	y
8	=A2	=IF(A8<$A$3,$M$11+$M$12*A8+$M$13*A8^2+$M$14*A8^3,$M$15+$M$16*A8+$M$17*A8^2+$M$18*A8^3)
9	=A8+0.5	=IF(A9<$A$3,$M$11+$M$12*A9+$M$13*A9^2+$M$14*A9^3,$M$15+$M$16*A9+$M$17*A9^2+$M$18*A9^3)

Figure 3.55

From the graph we see that the model forms a smooth curve that goes through each data point, as required.

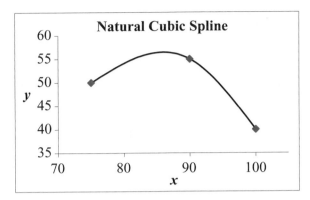

Figure 3.56

For large sets of data the calculations for a cubic spline model can get quite complicated. For 11 data points, the model is made up of 10 separate cubic polynomials with a total of 40 unknown parameters. This results in a 40×40 matrix A. Excel can handle up to a 52×52 matrix, but setting it up is quite tedious. Fortunately, there are optional add-ins (and other software) that do the work automatically.

Exercises

3.6.1 To get the necessary eight equations to determine the values of the eight parameters in our cubic spline model, we added the conditions that $p_1'' (x_1) = p_2'' (x_3) = 0$. Setting these second derivatives equal to 0 is somewhat arbitrary. Let's experiment with changing these conditions.

1. In the worksheet **Cubic**, change the condition $p_2'' (x_3) = 0$ to values other than 0 (i.e., change the number in cell **M9** to values other than 0). Does this change the shapes of the graphs of $p_1 (x)$ and $p_2 (x)$? (Use the data points shown in Figure 3.54 where $x_3 = 100$.)

2. In the original natural cubic spline model (with $p_2'' (x_3) = 0$), the model predicts a sharp decrease in milk production around day 100 (in other words, $p_2' (x_3)$ is a large negative number). In the graph of the original data shown in Figure 3.49, we see that the milk production begins to "level off" between days 100 and 120. Find a value of $p_2'' (x_3)$ that gives a model that "levels off" near $x = 100$.

3. Instead of specifying the value of $p_2'' (x_3)$ in the last row of matrix A, we could replace this with any condition on $p_1 (x)$ or $p_2 (x)$, or their derivatives, that is independent of the other conditions. Notice that in the linear spline model the slope between days 100 and 120 is -0.5. Modify the worksheet **Cubic** by replacing the last row of matrix A with the condition that $p_2' (x_3) = -0.5$. Does the resulting model "level off" near $x = 100$?

3.6.2 Modify the worksheet **Cubic** to calculate and graph a cubic spline model fit to *four* data points.

3.6.3 Consider the problem of fitting a *quadratic* spline model to a set of three data points $\{(x_1, y_1), (x_2, y_2), (x_3, y_3)\}$ where $x_1 < x_2 < x_3$. The two polynomials have the general form

$$p_1(x) = a_1 + b_1 x + c_1 x^2 \quad \text{for } x_1 \leq x < x_2$$
$$p_2(x) = a_2 + b_2 x + c_2 x^2 \quad \text{for } x_2 \leq x \leq x_3.$$

1. If these polynomials must satisfy the same three conditions as the cubic polynomials, find a set of six equations that determine the values of the parameters a_1, b_1, c_1, a_2, b_2, and c_2.

2. Design a spreadsheet to calculate the values of the parameters (use the same data points as in Example 3.6.2). What do you observe about the two polynomials $p_1(x)$ and $p_2(x)$?

3. Fit a second-degree polynomial trendline to the data. How does this polynomial compare to $p_1(x)$ and $p_2(x)$? Why is this?

3.6.4 A researcher observes a particle moving in a straight line and measures its velocity at three points in time as recorded in Table 3.14.

Time (s)	0	2	5
Velocity (m/s)	0	60	85

Table 3.14

The researcher wants to know how far the particle traveled over the interval of time $[0, 5]$. In general, if $f(t) = $ velocity of an object at time t, we can calculate the distance it traveled over an interval of time $[t_0, t_1]$ by

$$\text{distance} = \int_{t_0}^{t_1} |f(t)| dt.$$

1. Fit a linear spline model to the data in Table 3.14 to estimate $f(t)$ and use the result to estimate the distance traveled. What is the meaning of the slope of each of the line segments?

2. Fit a cubic spline model to the data in Table 3.14 to estimate $f(t)$ and use the result to estimate the distance traveled. Also, estimate the acceleration at time $t = 1.5$.

3. Which of the two estimates of distance traveled do you think is more accurate? Why?

For Further Reading

- For more information on the 10-year fluctuations of wildlife populations, see L. G. Keith. 1963. *Wildlife's ten year cycle.* Madison, WI: University of Wisconsin Press.

- For more information on least-squares solutions and their applications to linearizable models, see D. C. Lay. 2006. *Linear algebra and its applications.* 3rd ed. Boston: Pearson Addison Wesley, 409–425.

- For a much more detailed theoretical discussion of most of the topics discussed in this chapter, see W. W. Hines. 2003. *Probability and statistics in engineering.* 4th ed. Hoboken, NJ: John Wiley & Sons, 409–486.

References

Florida Fish and Wildlife Conservation Commission. Marine Mammal Pathobiology Laboratory. 2009. *http://research.myfwc.com/features/view_article.qsp?id=12084*

Gordon, S. 1999. Modeling the U.S. population. *AMATYC Review* 20(2), 17–29.

Sincich, T., D. M. Levine, and D. Stephan. 2002. *Practical statistics by example using Microsoft Excel and Minitab.* 2nd ed. Upper Saddle River, NJ: Prentice Hall, 602.

World almanac and book of facts. 1992. New York, NE: Pharos Books.

World almanac and book of facts. 2007. Mahwah, NJ: World Almanac Books.

CHAPTER 4

Discrete Dynamical Systems

Chapter Objectives

- Define and solve discrete dynamical systems
- Analyze the long-term behavior of discrete dynamical systems numerically and graphically
- Model different scenarios with linear and nonlinear discrete dynamical systems

4.1   Introduction

A *dynamical system* is simply a system that changes over time. The bacterial growth modeled in Chapter 1 is one such example. When time is measured in discrete units, such as in the bacterial growth example, the system is called a *discrete* dynamical system.

Dynamical systems are "easy to [model] and hard to solve" (Meerschaert 1999, 127). In this chapter we introduce basic techniques for formulating dynamical models and graphical approaches to their analysis.

In mathematical terms, a discrete dynamical system is simply a sequence of numbers. Consider a savings account that is compounded yearly and the interest is added at the end of each year. If a_n is the amount in the account at the end of year n ($n = 0, 1, \ldots$) and r is the interest rate, we have the sequence

$$a_1 = a_0 + r\,a_0 = (1 + r)\,a_0$$
$$a_2 = a_1 + r\,a_1 = (1 + r)\,a_1$$
$$\vdots$$
$$a_{n+1} = a_n + r\,a_n = (1 + r)\,a_n.$$

This last equation leads us to the formal definition of a dynamical system.

Definition 4.1.1 A *discrete dynamical system* is a sequence of numbers $\{\,a_n \mid n = 0, 1, \ldots\}$ defined by a relation of the form

$$a_{n+1} = f\,(a_n)$$

where f is some real-valued function.

The variable a_n is generically referred to as the *state of the system*. In simpler terms, a discrete dynamical system is one in which the state of the system is determined by the previous state. In the savings account example there is only one component to the system (the amount in the account), so it is called *one-dimensional*.

When the function f has the form $f(x) = bx$ where b is a constant, the dynamical system is referred to as *linear*.

Definition 4.1.2 A *linear discrete dynamical system* is a sequence of numbers $\{\,a_n \mid n = 0, 1, \ldots\}$ defined by a relation of the form

$$a_{n+1} = ba_n$$

where $b \neq 0$ is a constant.

Any dynamical system that does not have this linear form is generically referred to as *nonlinear*. A *solution* of a discrete dynamical system is an explicit description of a_n in terms of n and the initial state a_0.

Theorem 4.1.1 *The solution of a linear dynamical system $a_{n+1} = ba_n$ for $b \neq 0$ is*

$$a_n = b^n a_0 \tag{4.1}$$

where a_0 is the initial state.

Proof: We first need to show that (4.1) satisfies the initial condition. Note that in (4.1),

$$a_0 = b^0 a_0 = a_0,$$

so the initial condition is satisfied. Next, we need to show that (4.1) satisfies the definition of a linear discrete dynamical system. Note that

$$a_{n+1} = b^{n+1} a_0 = b(b^n a_0) = ba_n$$

as required.   □

4.2 Long-Term Behavior and Equilibria

Theorem 4.1.1 gives us an easy-to-use formula for finding the exact value of a_n. However, we are usually more interested in describing the long-term behavior of the system than in finding exact values at points in time. That is, we want to know what happens to a_n for large values of n.

Example 4.2.1   Long-Term Behavior

Let's graphically examine the long-term behavior of a linear dynamical system $a_{n+1} = ba_n$ for various values of b. For simplicity, suppose that $a_0 = 0.1$.

1. Rename a blank worksheet "**Linear**" and format it to look like Figure 4.1. Copy the formulas in **A3:B3** down to row 17. This will give the first 16 values of a_n ($0 \leq n \leq 15$) with $b = 0.5$.

2. Highlight the column titled **n** and **a_n** and create a graph similar to Figure 4.2. Set the **x-axis min** and **max** to 0 and 15, respectively. Notice that for this value of b, the state of the system approaches 0 as n gets larger.

	A	B	C
1	n	a_n	b
2	0	0.1	0.5
3	=A2+1	=B2*$C$2	

Figure 4.1

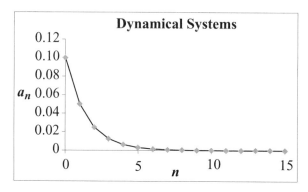

Figure 4.2

3. Next open the **Control** toolbox by selecting **View** → **Toolbars** → **Control Tool-box**. Draw a scroll bar by selecting the icon on the left side of the **Control Toolbox** window that is second from the bottom. Your cursor will turn into a cross. Draw a horizontal scroll bar near the top of the worksheet (see Appendix A.4 for more information on drawing scroll bars).

4. Right-click on the scroll bar and select **Properties**. Set the **LinkedCell** to **D5**, the **max** to 100, and the **min** to 0. Close the **Properties** window. Exit **Design Mode** by selecting the icon in the upper-left corner of the **Control Toolbox**.

5. Slide the scroll bar left and right. The number in **D5** should change between 0 and 100. Enter a formula in **C2** as shown in Figure 4.3. This will allow us to change the value of b between -2 and $+2$ with the scroll bar.

	C
1	b
2	=-2+0.04*D5

Figure 4.3

6. Move the slider on the scroll bar left and right and notice how the behavior of the system changes, especially when b passes -1, 0, and $+1$. Our observations are

summarized in Table 4.1. As we can see, the long-term behavior of the system is dramatically affected by the value of b.

Value of b	Behavior of a_n
$b < -1$	Oscillates between positive and negative; $|a_n|$ grows without bound
$b = -1$	Oscillates between a_0 and $+a_0$
$-1 < b < 0$	Oscillates between positive and negative; $|a_n|$ approaches 0
$b = 0$	$a_n = 0$ for $n > 0$
$0 < b < 1$	a_n approaches 0
$b = 1$	$a_n = a_0$ for all n
$b > 1$	a_n grows without bound

Table 4.1

Now consider a savings account that pays 5% interest compounded yearly. We saw in Section 4.1 that an account with an interest rate r is modeled by

$$a_{n+1} = (1 + r)\, a_n.$$

In this case, we have $r = 0.05$, so our model is

$$a_{n+1} = 1.05\, a_n.$$

Suppose now that we want to withdraw $2000 at the end of each year to supplement our income. We want to know how much money we need to deposit now so that we never run out of money.

To answer this question, we will analyze a slightly more general problem: What happens to the amount in the account in terms of the initial deposit? First we construct our model. The amount in the account grows at 5% compounded yearly but we are withdrawing $2000 each year. A dynamic model that describes this scenario is

$$a_{n+1} = 1.05\, a_n - 2000.$$

As before, a_n is the amount in the account at the end of year n. We are also assuming that there is no penalty for withdrawing money each year and that we withdraw the money after the interest from the previous year has been added. This system is an example of an *affine* dynamical system.

Definition 4.2.1 An *affine discrete dynamical system* is a sequence of numbers $\{\, a_n \,|\, n = 0, 1, \ldots\}$ described by a relation of the form

$$a_{n+1} = b a_n + m$$

where $b \neq 0$.

Central to the analysis of the long-term behavior of any dynamical system are *equilibrium values* (also called *fixed points*).

Definition 4.2.2 A number a is called an *equilibrium value* for the dynamical system $a_{n+1} = f(a_n)$ if $a_n = a$ for all n whenever $a_0 = a$.

To find equilibrium values, note that if a is an equilibrium value, we must have

$$a_{n+1} = a_n = a \quad \Rightarrow \quad f(a) = a.$$

So, finding equilibrium values simply requires us to solve the equation $f(a) = a$. For an affine system, we have

$$a = ba + m \quad \Rightarrow \quad a = \frac{m}{1 - b}.$$

In this example, $b = 1.05$ and $m = -2000$, so the equilibrium value is $a = \frac{-2000}{1-1.05} = 40,000$. Thus, if we start with \$40,000 in the account and withdraw \$2000 at the end of each year, we will always have the same amount in the account at the end of each year.

Example 4.2.2   Savings Account

We will take a graphical approach to analyze what happens for initial values other than the equilibrium value of \$40,000.

1. Rename a blank worksheet "**Savings**" and format it as in Figure 4.4. Copy the range **A3:B3** down to row 27 to model the account over the first 25 years.

	A	B	C	D
1	**n**	**a$_n$**	**r**	**m**
2	0	40000	0.05	2000
3	=A2+1	=(1+$C$2)*B2–$D$2		

Figure 4.4

2. Use the data in columns **A** and **B** to form a graph as in Figure 4.5. Set the **y-axis min** and **max** to **0** and **80,000**, respectively, and the **x-axis min** and **max** to **0** and **25**, respectively. Note that the value in the account does not change if we start with \$40,000, as expected.

3. Next, add a scroll bar. Set the linked cell to **B2** and the **min** and **max** to **0** and **80,000**, respectively. This will allow us to vary the value of a_0 between \$0 and \$80,000 in increments of \$1.

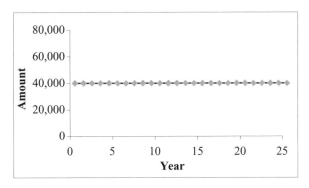

Figure 4.5

4. Move the slider on the scroll bar left and right and observe how the long-term behavior of the system changes. Specifically, note that for a_0 less than \$40,000, the amount in the account eventually decreases to 0 and for a_0 greater than \$40,000, the amount grows without bound.

In Example 4.2.2 we saw that the long-term behavior of the system changed quite dramatically with a small change in a_0. In situations like this we say that the system is *sensitive to the initial condition.*

Also note that if $a_0 \neq 40,000$, the system either approaches 0 or increases without bound. The equilibrium value of 40,000 is an example of an *unstable* or *repelling* equilibrium.

Definition 4.2.3 An equilibrium value a is *unstable* or *repelling* if there is a number ε such that

$$|a_n - a| > \varepsilon \quad \text{whenever} \quad |a_0 - a| < \varepsilon$$

for some n. The equilibrium a is *stable* or *attracting* if there is a number ε such that

$$\lim_{n \to \infty} a_n = a \quad \text{whenever} \quad |a_0 - a| < \varepsilon.$$

In less technical terms, a is unstable if the system starts near it, but does not approach it. This is exactly what we saw in Example 4.2.2. The equilibrium is stable if the system starts near a and approaches it.

Example 4.2.3   Antibiotic in the Bloodstream

An infant is given the antibiotic Cefdinir to treat an ear infection. When taking an antibiotic, it is important to keep the amount of the drug in the bloodstream fairly constant. If the amount gets too low, the bacteria can begin to regrow. If the amount gets too high, it could cause other complications.

Suppose the half-life of the drug is 1 day (meaning that half the Cefdinir remains in the blood after each 1-day period) and a dosage of 0.1 mg is given at the end of each day. Let's examine what happens to the amount of Cefdinir in the bloodstream in the long run.

A simple affine model for this system is

$$a_{n+1} = 0.5\,a_n + 0.1$$

where a_n = the amount of Cefdinir in the blood at the end of day n. Since the problem did not specify the initial dosage, a_0, we need to experiment with different values.

1. Rename a blank worksheet "**Cefdinir**" and format it as in Figure 4.6. Copy row 3 down to row 17 to model the system from day 0 to day 15.

	A	B
1	n	a_n
2	0	0
3	=A2+1	=0.5*B2+0.1

Figure 4.6

2. Create a graph similar to that in Figure 4.7. Notice that even with an initial dosage of 0 mg, the amount of Cefdinir in the blood appears to approach 0.2 mg at the end of each day. Note that this does not mean that the amount eventually equals 0.2 mg at every point in time, only that it equals 0.2 mg at the end of every day.

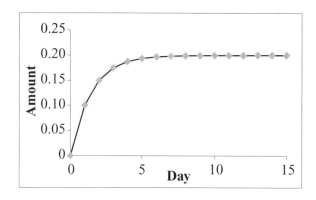

Figure 4.7

3. Next, add a scroll bar and set the **min** to **0**, the **max** to **100**, and the linked cell to **C1**. Add the formula in Figure 4.8 to allow us to vary the initial dosage from 0 to 1 mg in increments of 0.01 mg.

	B
1	a_n
2	=C1/100

Figure 4.8

4. Move the slider on the scroll bar left and right and observe the long-term behavior of the system. Specifically, note that when $a_0 = 0.2$, the system remains at 0.2, meaning that 0.2 is an equilibrium value. Also note that no matter what the value of a_0, the system appears to always approach 0.2. This shows that 0.2 is a *stable equilibrium*.

Example 4.2.4   Generalized Savings Account

Let's generalize the model of the savings account. Suppose that we want to withdraw $5000 every other year (starting in year 2) instead of every year. This makes our model

$$a_{n+1} = \begin{cases} 1.05a_n - 5000 & \text{if } n+1 \text{ is a multiple of } 2 \\ 1.05a_n & \text{otherwise} \end{cases}$$

This system is neither linear nor affine. We will take a strictly graphical approach to analyze the behavior in terms of the initial deposit.

1. Modify the worksheet "**Savings**" as in Figure 4.9 and copy the range **B3:C3** down to row 27. The **MOD** function in cell **C3** returns the remainder when the year n (cell **A3**) is divided by the number of years between withdrawals (cell **F3**). If this remainder is 0, then the year is a multiple of the years between withdrawals, and a withdrawal is taken that year. Otherwise, no withdrawal is taken. Note also that we have designed a cell to hold the parameter "Years between withdrawals" so that we can easily change its value to analyze other scenarios.

	A	B	C	D	E	F
1	n	a_n	Withdrawal?	r	Amount of	Years between
2	0	37802	0	0.05	withdrawal	withdrawals
3	=A2+1	=(1+$D$2)*B2–C3*$E$3	=IF(MOD(A3,$F$3)=0,1,0)		5000	2

Figure 4.9

2. Move the slider on the scroll bar left and right to change the value of a_0 and observe the long-term behavior of the system. Specifically, note that for a_0 below approximately $48,000, the account eventually is depleted. For a_0 above $48,000, the value tends to increase. For a_0 around $48,000, the value fluctuates around $48,000.

Exercises

4.2.1 Suppose you open a savings account that pays 5% interest compounded yearly with a $500 initial deposit and make a $200 deposit at the end of each year. Construct a model of the amount of money in the account at the end of each year and define the variables.

4.2.2 Consider the linear dynamical system $a_{n+1} = ba_n$.

1. Suppose $b = 1$. Describe the equilibrium values of the system. Are they stable or unstable?

2. Suppose $b \neq 1$. Describe the equilibrium value of the system. For which values of b is this equilibrium value stable? For which is it unstable?

4.2.3 Analytically (meaning don't use a spreadsheet), show that $a_n = b^n \left(a_0 - \frac{m}{1-b} \right) + \frac{m}{1-b}$ is the solution to the affine system $a_{n+1} = ba_n + m$ where $b \neq 0$.

4.2.4 Use the solution to Exercise 4.2.3 to find the amount of time it would take the account in Exercise 4.2.1 to build a value of $12,000.

4.2.5 Your grandparents have their life savings of $750,000 in a savings account that pays 6.7% interest compounded yearly. They want to spend all of it before they die. If they plan to live 15 more years, how much should they withdraw at the end of each year to accomplish their goal?

4.2.6 The amount of a drug in a patient's bloodstream decreases at the rate of 50% per hour. Suppose an injection is given at the end of each hour that increases the amount of drug in the bloodstream by 0.2 unit.

1. Formulate a model of the amount of drug in the bloodstream at the end of each hour.

2. Find the equilibrium value(s) of your model.

3. Graphically, classify each equilibrium value as stable or unstable.

4. Suppose we give the injection every 3 h. Describe what happens to the long-term level of the drug in the bloodstream.

4.2.7 In the generalized model of the savings account where we withdraw $5000 every 2 years, find a value of a_0 such that $a_2 = a_0$. If the initial deposit is this value, what is the long-term behavior of the system?

4.2.8 Suppose two countries are engaged in an arms race. Further suppose that the two countries have economies of similar strength and they have similar levels of distrust of each other. A simple model for T_n, the total amount of money spent by the two nations, is given by the affine system

$$T_{n+1} = (1 - r + d) T_n + c$$

where r is a positive constant that measures the restraint of growth due to the strength (or weakness) of the economies of the countries, d is a positive constant that measures the level of distrust between the countries, and c is a constant. If T_n eventually grows too large, then the countries will not be able to support the arms race and either they must negotiate an end or war will break out. If T_n approaches a constant level, then they have a "stable" arms race. If T_n eventually decreases to 0, then the race ends.

1. Use the solution in Exercise 4.2.3 to find the solution to this affine system.

2. If $-1 < (1 - r + d) < 1$, use your solution in part (1) to find $\lim\limits_{n \to \infty} T_n$. What happens to the arms race in this situation? Since $(1 - r + d) < 1$, what relationship is there between d and r?

3. If $T_0 = 199$, $c = -127$, $r = 1/3$, and $d = 1$, use your solution in part (1) to find $\lim\limits_{n \to \infty} T_n$. What happens to the arms race in this situation?

4. Now suppose that $T_0 = 181$, $c = -127$, $r = 1/3$, and $d = 1$. Find $\lim\limits_{n \to \infty} T_n$. What happens to the arms race in this situation?

5. Generalize your findings in the last two parts. Suppose that $(1 - r - d) > 1$ and find a relationship between T_0 and the fraction $\frac{c}{r-d}$ so that the arms race dies out. What is the relationship that leads to war or negotiations?

4.3   Growth of a Bacteria Population

Table 4.2 gives the number of bacteria in a Petri dish, a_n, at the end of each hour n. This data is graphed in Figure 4.10. We want to model a_n in terms of n.

n	0	1	2	3	4	5	6	7	8	9
a_n	10.3	17.2	27	45.3	80.2	125.3	176.2	255.6	330.8	390.4
n	10	11	12	13	14	15	16	17	18	19
a_n	440	520.4	560.4	600.5	610.8	614.5	618.3	619.5	620	621

Table 4.2

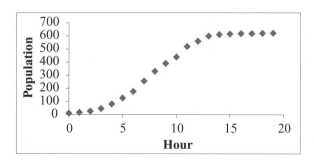

Figure 4.10

When modeling a dynamical system, it is often convenient to think about the way the variable(s) change(s) between time periods. As in Chapter 2, we consider the change between time periods $\Delta a_n = a_{n+1} - a_n$. The values of this variable for the first eight values of n are given in Table 4.3.

n	0	1	2	3	4	5	6
a_n	10.3	17.2	27	45.3	80.2	125.3	176.2
Δa_n	6.9	9.8	18.3	34.9	45.1	50.9	79.4

Table 4.3

Notice that as a_n increases, Δa_n also increases. This suggests that Δa_n is proportional to a_n, which leads to the equation

$$\Delta a_n = a_{n+1} - a_n = r\,a_n \tag{4.2}$$

where r is some positive constant. An equation describing the difference in populations between time periods, such as (4.2), is called a *difference equation*. Forming a difference equation is often the first step in modeling a discrete dynamical system. Solving this equation for a_{n+1} yields the model

$$a_{n+1} = (1+r)\,a_n.$$

The parameter r can be interpreted as a constant hourly growth rate. However, the graph of population versus hour shows that the population does not grow at a constant rate. Also note that this constant hourly growth rate would predict a population that grows without bound, which the data does not support either.

To refine the model, note that the graph shows that the rate of growth decreases as the population nears 621. This number is called the *carrying capacity* of the system. So instead of assuming a constant growth rate r, we assume a growth rate that approaches zero as the population approaches 621. An equation implementing this assumption is

$$\Delta a_n = a_{n+1} - a_n = b(621 - a_n)\,a_n \tag{4.3}$$

where $b > 0$ is a constant. Again, solving for a_{n+1} yields the model

$$a_{n+1} = a_n + b(621 - a_n)\,a_n. \tag{4.4}$$

Equation (4.4) is an example of a *discrete logistic equation*.

Definition 4.3.1 A *discrete logistic equation* (also called a *logistic map* or a *constrained growth model*) is an equation of the form

$$a_{n+1} = a_n + b\,(c - a_n)\,a_n$$

where b and c are constants. This type of equation is often used to model population growth where a_n is the population at time n. The constant b is called the *intrinsic growth rate* and c is called the *carrying capacity*.

Example 4.3.1   Bacteria Growth

To implement the model (4.4) we need to find the value of b. Equation (4.3) predicts that $(a_{n+1} - a_n)$ is proportional to $(621 - a_n) a_n$. If a graph of $(a_{n+1} - a_n)$ versus $(621 - a_n) a_n$ is approximately a straight line through the origin, then the assumption is reasonable and the slope of the line is the value of b.

1. Rename a blank worksheet "**Bacteria**" and format it as in Figure 4.11. Enter the data from Table 4.2 in columns **A** and **B** and copy range **D2:E2** down to row 20.

	A	B	C	D	E
1	n	a_n	Predicted	$a_n(621-a_n)$	$a_{n+1}-a_n$
2	0	10.3		=B2*(621−B2)	=B3−B2

Figure 4.11

2. Create a graph of the transformed data in columns **D** and **E** and fit a straight line through the origin as in Figure 4.12. We see that the line fits the data well, so our model appears to be reasonable.

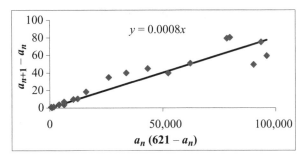

Figure 4.12

3. Using the slope of the line in Figure 4.12, our model is

$$a_{a+1} = a_n + 0.0008 \left(621 - a_n\right) a_n.$$

To test this model against the given data add the formula in Figure 4.13 and copy row 3 down to row 21.

	C
1	Predicted
2	10.3
3	=C2 + 0.0008* (621−C2)*C2

Figure 4.13

4. Use the data in columns **A**, **B**, and **C** to form a graph as in Figure 4.14. Notice that the predicted values are relatively close to the actual values, so our model is verified.

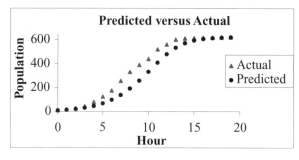

Figure 4.14

Example 4.3.2   Sensitivity to the Intrinsic Growth Rate

Earlier we looked at how the long-term behavior of the savings account changes as the initial deposit changes. Now we look at how the behavior of our constrained population model changes as the value of b changes. We take a strictly graphical approach.

1. Rename a blank worksheet "**Bacteria 2**" and format it as in Figure 4.15. Copy row 3 down to row 100.

	A	B	C
1	n	a_n	b
2	0	10.3	0.0008
3	=A2+1	=B2 + $C$2*(621−B2)*B2	

Figure 4.15

2. Highlight columns **A** and **B** and form a graph as in Figure 4.16. Set the **y-axis min** and **max** to **0** and **800**, respectively.

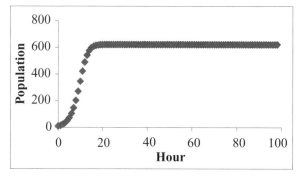

Figure 4.16

3. Next add a scroll bar, set the linked cell to **D2**, set the **min** and **max** to **0** and **900**, respectively, and add the formula in Figure 4.17. This will allow us to vary the value of b between 0 and 0.0045.

	C
1	**b**
2	=D2/200000

Figure 4.17

4. Move the slider left and right and notice how the behavior of the system changes, especially when b increases above 0.0026. Several examples are shown in Figure 4.18.

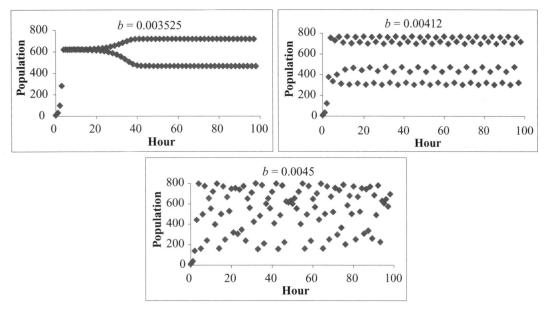

Figure 4.18

Notice that the behavior of the system changes quite dramatically as b changes. For small values of b, the behavior is very regular. But for larger values, it is quite chaotic. This illustrates that we must be careful about using a model of this form if we are unsure of the value of b. The value of b dramatically affects the behavior of the system. If we use a wrong value of b, our analysis of the system could be very inaccurate.

Exercises

4.3.1 Table 4.4 contains data on the population of foxes in a forest over a period of several years. Fit a discrete logistic equation to the data. How well does the model fit the data?

n	0	1	2	3	4	5	6	7	8	9	10
a_n	50	85	110	130	175	200	215	221	228	232	234

Table 4.4

4.3.2 In this exercise we will fit a discrete logistic equation to the data in Table 4.2 using the least-squares criterion. This criterion says, informally, that when predicting the values of data $\{a_n \mid n = 1, \dots, m\}$, the number

$$\sum_{i=1}^{m} (a_n - \text{Predicted})^2$$

(called the sum of squares) should be as small as possible. With a carrying capacity of 621, use a scroll bar to find a value of b in the discrete logistic equation that minimizes the sum of squares. Does this give the same model as the one found in Example 4.3.1?

Here are some suggestions:

1. Modify the worksheet **Bacteria** by adding a column titled "$(a_n - \text{Predicted})^2$." Sum the values in this column to calculate the sum of squares.

2. Create a cell to hold the value of b. Reference this cell in your formula for the predicted values.

3. Add a scroll bar with a **min** and **max** of **0** and **1000**, respectively. Create a formula for the value of b equal to the scroll bar–linked cell divided by 500,000.

4. Move the slider back and forth to find a value of b that minimizes the sum of squares.

4.3.3 Suppose we estimate that a forest can support a population of 10,000 deer and that the population of deer is described by the model $a_{n+1} = a_n + 0.00006(10000 - a_n) a_n$ where a_n is the population at the end of year n.

1. Suppose that we let hunters kill 700 deer at the end of each year. Write a model to describe this situation and analyze the long-term behavior of the population for different initial populations.

2. Suppose we start with a population of 9,000 deer and we allow hunters to kill m deer at the end of each year. Analyze the long-term behavior of the population for different values of m. What is the maximum value of m for which the population survives in the long-term?

3. Suppose we start with a population of 9000 deer and we allow hunters to kill a certain proportion of the population (such as 0.25) at the end of each year. Write a model to describe this situation and analyze the long-term behavior of the population for different values of the proportion.

4.4   A Linear Predator–Prey Model

Consider a forest containing foxes and rabbits where the foxes eat the rabbits for food. Let's examine whether the two species can survive in the long term. A forest is a very complex ecosystem. So to simplify the model, we use the following assumptions:

1. The only source of food for the foxes is rabbits and the only predator of the rabbits is foxes.
2. Without rabbits present, foxes would die out.
3. Without foxes present, the population of rabbits would grow.
4. The presence of rabbits increases the rate at which the population of foxes grows.
5. The presence of foxes decreases the rate at which the population of rabbits grows.

We will model these populations using a discrete dynamical model. Each state of the system consists of the populations of foxes and rabbits at a point in time. Since this state consists of two components, this is a *two-dimensional discrete dynamical system*.

To create our model, we first need to define some variables. Let

$$F_n = \text{population of foxes at the end of month } n$$
$$R_n = \text{population of rabbits at the end of month } n.$$

As in the bacteria model, the assumptions are stated in terms of rates of change, $\Delta F_n = F_{n+1} - F_n$ and $\Delta R_n = R_{n+1} - R_n$. There are many ways we could model these rates of change with the assumptions. In this section we will create a linear model. In the next section we will create a nonlinear model.

Assumptions (2) and (3) deal with the rates of change of each population in the absence of the other. A reasonable way to model these is to say that the rates are proportional to the populations. This yields the difference equations

$$\Delta F_n = F_{n+1} - F_n = -a\,F_n \tag{4.5}$$
$$\Delta R_n = R_{n+1} - R_n = d\,R_n \tag{4.6}$$

where $0 < a, d \le 1$. Note that the coefficient of proportionality in equation (4.5) is negative to reflect the fact that the foxes would die out (a negative rate of change) without the rabbits. The coefficient in equation (4.6) is positive because the population of rabbits grows (a positive rate of change) without the foxes.

Now, assumptions (4) and (5) say that these rates in equations (4.5) and (4.6) either increase or decrease in the presence of the other species. So to incorporate these assumptions, we simply add one term to each of equations (4.5) and (4.6), yielding

$$F_{n+1} - F_n = -a\,F_n + bR_n \tag{4.7}$$
$$R_{n+1} - R_n = -cF_n + d\,R_n \tag{4.8}$$

where $0 \leq b,\ c$. Note that the added term in equation (4.7) is positive to reflect the fact that the presence of rabbits increases the rate at which the population of foxes grows. The added term in equation (4.8) is negative because the presence of foxes decreases the rate at which rabbits grow.

Rewriting equations (4.7) and (4.8) yields our model in the form of a system of linear equations:

$$F_{n+1} = (1 - a)F_n + bR_n \tag{4.9}$$
$$R_{n+1} = -cF_n + (1 + d)R_n. \tag{4.10}$$

Because our model has the form of a system of linear equation, it is called a *two-dimensional linear discrete dynamical system.*

The model could be written in matrix form as

$$\begin{bmatrix} F_{n+1} \\ R_{n+1} \end{bmatrix} = \begin{bmatrix} 1-a & b \\ -c & 1+d \end{bmatrix} \begin{bmatrix} F_n \\ R_n \end{bmatrix}. \tag{4.11}$$

For a description of how to analyze this model using matrix and eigenvalue techniques, see, for instance, *Linear Algebra and Its Applications* (Lay 2003). We will take a strictly graphical approach to analyze the model.

The parameters $(1 - a)$ and b will generically be called the fox death and birth factors, respectively, while the parameters $-c$ and $(1 + d)$ will be called the rabbit death and birth factors, respectively.

1. Rename a blank worksheet "**Linear Predator–Prey**" and format it as in Figure 4.19. The initial values of the parameters and populations are shown in the figure. Copy row 8 down to row 37 to model 30 months.

2. Next, add the graphs in Figure 4.20. The curve formed in the graph of rabbits versus foxes is called a *trajectory* of the system. The $x - y$ plane is called the *phase plane.* Notice that the curve tends toward the origin (0 foxes and 0 rabbits). This means that both species eventually die out. This is also shown in the other two graphs. Change the initial populations and observe that the trajectories also tend toward the origin.

	A	B	C
1		**Factors**	
2		**Death**	**Birth**
3	**Foxes**	0.5	0.4
4	**Rabbits**	-0.17	1.1
5			
6	**Month**	**Foxes**	**Rabbits**
7	0	500	200
8	=A7+1	=$B$3*B7+C7*$C$3	=B7*$B$4+C7*$C$4

Figure 4.19

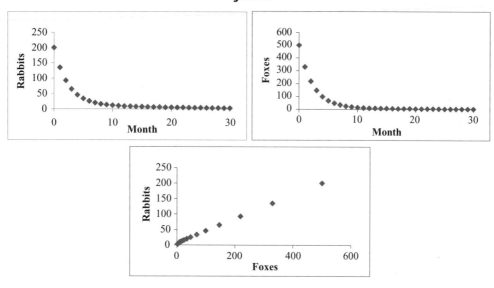

Figure 4.20

As in a one-dimensional discrete dyanmical system, two-dimensional systems can have an *equilibrium*.

Definition 4.4.1 Let

$$a_{n+1} = f(a_n, b_n), \; b_{n+1} = g(a_n, b_n)$$

be a two-dimensional discrete dynamical system. A point (a, b) is an *equilibrium* if $a_n = a$ and $b_n = b$ for all n whenever $a_0 = a$ and $b_0 = b$.

Stated another way, (a, b) is an equilibrium point if $f(a, b) = a$ and $g(a, b) = b$. For the system given by equations (4.9) and (4.10), the origin $(0, 0)$ is an equilibrium point because

$$f(0, 0) = (1 - a)0 + b0 = 0 + 0 = 0$$
$$g(0, 0) = -c0 + (1 + d)0 = 0 + 0 = 0.$$

Since the trajectories appear to be attracted to the origin, we have graphical evidence that the origin is a *stable equilibrium*.

Example 4.4.1   Bifurcation

Let's examine what happens if the rabbit death rate changes.

1. Add a scroll bar to the worksheet; set the **min** and **max** to **0** and **500**, respectively, and the linked cell to **H1**.

2. Add the formula in Figure 4.21 to vary the rabbit death rate from 0 to -0.50 in increments of 0.001.

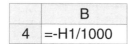

	B
4	=-H1/1000

Figure 4.21

3. Move the slider on the scroll bar left and right and notice how the behavior of the system changes. For values of the death rate less than -0.123, the origin appears to be stable and both populations die out, as in Figure 4.20. For values greater than -0.123, the origin appears to be unstable and both populations grow without bound, as in Figure 4.22. This change of behavior caused by a change in the value of a parameter is called a *bifurcation*.

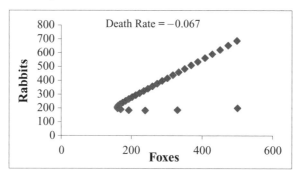

Figure 4.22

Notice that this model predicts that both populations either grow without bound or die off. These are two extremes. The model in Section 4.5 is a refinement of this model that allows for other possibilities.

Exercises

4.4.1 Suppose the rabbit death factor is -0.05 and the other factors are as shown in Figure 4.19. For initial populations of 500 foxes and 200 rabbits, we saw that the populations grow without bound in the long run. Here we investigate whether this is true for all initial populations.

1. Fix the initial rabbit population at 200. What happens if the initial fox population increases or decreases? How large can the initial fox population be for both species to survive?

2. Fix the initial fox population at 500. What happens if the initial rabbit population increases or decreases? How small can the initial rabbit population be for both species to survive?

4.4.2 Investigate sensitivity to the parameter "rabbit birth factor" using values of the other parameters and the initial populations shown in Figure 4.19. This means to change the value of the rabbit birth factor in small amounts up and down and note any changes in the behavior of the system. A forest ranger knows that the populations will both die out under the parameters in Figure 4.19. To prevent this, she advocates building rabbit shelters in an attempt to increase the survival rate of baby rabbits. Does this model predict that this change would have any effect on the survival of the populations? Why or why not?

4.4.3 With the parameters and the initial populations shown in Figure 4.19 both populations die out. A hunting club claims that if its members are allowed to kill (or harvest) a few foxes each month the populations would survive.

1. Modify the model to include this harvesting. Does the model support this claim? If it does, how many foxes could be harvested each month for both species to survive?

2. A conservation group claims that more foxes should be put into the forest each month for the populations to survive. Does your model support this claim?

4.4.4 A biologist models the populations of foxes and rabbits in another forest as follows:

$$F_{n+1} = 1.3F_n + 0.4R_n$$
$$R_{n+1} = -0.6F_n + 1.05R_n.$$

1. Is this model consistent with the assumptions used in building our model? Why or why not?

2. What does this model suggest about the growth or death of foxes in the absence of rabbits?

4.5   A Nonlinear Predator–Prey Model

Let's consider a similar population of foxes and rabbits along with the same set of assumptions as in Section 4.4, but we will model the assumptions differently. We start with modeling assumptions (2) and (3) the same way:

$$\Delta F_n = F_{n+1} - F_n = -a\,F_n \tag{4.12}$$
$$\Delta R_n = R_{n+1} - R_n = d\,R_n \tag{4.13}$$

where $0 < a, d \le 1$. In Section 4.4, the coefficients of proportionality were kept constant. In this section we model them as increasing or decreasing in the presence of the other population, as we did in modeling the growth of bacteria.

Assumption (4) says that the presence of rabbits increases the rate of growth of foxes, so we have

$$F_{n+1} - F_n = (-a + bR_n)F_n \tag{4.14}$$

where $b \geq 0$. Likewise, assumption (5) says that the presence of foxes decreases the rate of growth of rabbits, so we have

$$R_{n+1} - R_n = (d - cF_n)R_n \tag{4.15}$$

where $c \geq 0$. Rewriting equations (4.14) and (4.15) we get our model:

$$F_{n+1} = (1 - a)F_n + bR_nF_n \tag{4.16}$$
$$R_{n+1} = -cR_nF_n + (1 + d)R_n. \tag{4.17}$$

This type of model is called a *Lotka–Volterra* model, named after the researchers who first devised it in the 1920s and 1930s.

Note that both equations have a term involving R_nF_n; thus the model is nonlinear. This term can be interpreted as modeling the number of *interactions* of the two species. These interactions increase the number of foxes while decreasing the number of rabbits. Also note the similarities between this nonlinear model and the linear model in equation (4.9). We refer to the parameters in this model using the same names as in the linear model.

This model can easily be implemented in Excel.

1. Rename a blank worksheet "**Nonlinear Predator–Prey**" and format it as in Figure 4.23. Copy row 8 down to row 507 to model 500 months. (Note that the parameters in this model have *similar* meanings as in the linear model, but they have different values. Also, we have different initial populations.)

	A	B	C
1		Factors	
2		Death	Birth
3	Foxes	0.88	0.0001
4	Rabbits	-0.0003	1.039
5			
6	Month	Foxes	Rabbits
7	0	110	900
8	= A7+1	= $B$3*B7+$C$3*B7*C7	= $B$4*B7*C7+$C$4*C7

Figure 4.23

2. Create graphs similar to those in Figure 4.24.

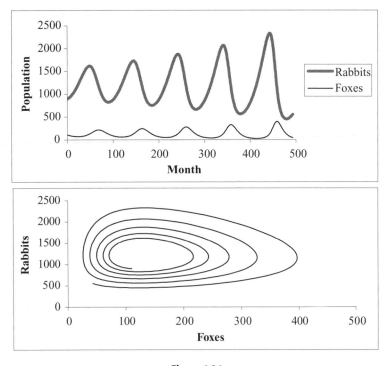

Figure 4.24

This model predicts that the populations oscillate with the same period of oscillation, but with a phase shift, meaning they do not reach their peaks at the same time. These oscillations cause the "spiral" nature of the trajectories in the graph of rabbits versus foxes. Oscillations such as this are actually observed in nature; thus this model appears to be more reasonable than the linear model.

Now let's calculate the equilibrium of the system. Suppose (f, r) is an equilibrium. By definition, this must satisfy the system of equations

$$f = 0.88f + 0.0001fr$$
$$r = -0.0003fr + 1.039r.$$

Assuming that $f \neq 0 \neq r$ yields the solution $f = 130$ and $r = 1200$. Another equilibrium is $(0, 0)$. Note that the point $(130, 1200)$ is at the "center" of the spiral in the phase plane. If we change the starting populations in the worksheet to 130 foxes and 1200 rabbits, we note that the populations do not change, as expected.

To determine whether this equilibrium is stable or unstable, we need to consider starting populations near the equilibrium. Changing the initial populations to 129 foxes and 1201 rabbits yields the trajectory shown in Figure 4.25. Notice that the trajectory moves away from the equilibrium value. Trying other initial populations yields similar results. The fact that the trajectories move away from the equilibrium is evidence that the equilibrium is unstable.

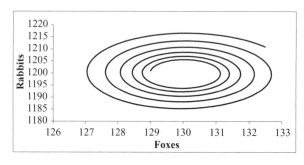

Figure 4.25

Example 4.5.1   Hunting

Suppose that a local hunting club wants to have a fox hunt at one of its next two meetings. The first meeting is 36 months from now and the next is 36 months after that. If the hunters limit themselves to killing 50 foxes, how would the two options affect the long-term populations of the foxes and rabbits?

Starting with 110 foxes and 900 rabbits, the model predicts that at month 36, there will be approximately 88 foxes. Killing 50 foxes would reduce this number to 38. In the worksheet, changing the number of foxes in month 36 to 38 results in the graph shown in Figure 4.26. Notice that this causes a dramatic change in the behavior of the system. The populations fluctuate much more than they did in Figure 4.24. Therefore, hunting 50 foxes in month 36 would have a great effect on the long-term populations.

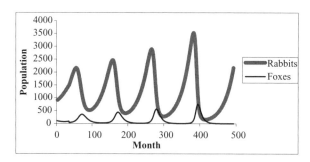

Figure 4.26

Now change the number of foxes in month 36 back to the original formula. In month 72, the model predicts a population of 212. Killing 50 foxes would leave 162. Changing the number of foxes in month 72 to 162 results in the graph shown in Figure 4.27. Notice that the fox population drops in month 72, which causes the rabbit population to initially grow. However, in the long term the populations behave much as they did in Figure 4.24.

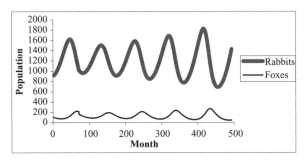

Figure 4.27

Why is there such a dramatic difference between these two cases? Note that originally near month 36, the fox population was near a local minimum. At month 72, it was near a local maximum. Killing 50 foxes near a time of a local minimum has a much greater effect than doing 80 near a local maximum.

Thus the hunting club should not schedule the hunt at month 36, as this would cause too great an effect on the populations. The hunt at month 72 would not cause a dramatic long-term effect.

Exercises

4.5.1 Consider the parameter "fox death factor."

1. Investigate the sensitivity of the system to the value of this parameter (use initial populations of 110 foxes and 900 rabbits).

2. Can it ever be greater than or equal to 1? Why or why not? (**Hint:** See equations (4.12) and (4.16).)

3. This parameter could be interpreted as the proportion of foxes that survive from one month to the next in the absence of rabbits. Suppose that a disease infects the fox population, decreasing the proportion that survives each month. What effect might this have on the populations?

4.5.2 Consider a model of the form

$$F_{n+1} = aF_n + bF_nR_n$$
$$R_{n+1} = cR_n + dF_nR_n$$

where a, b, c, and d are nonzero constants. Find a formula for the equilibria of the system in terms of a, b, c, and d.

4.5.3 Suppose hunters are allowed to kill m rabbits at the end of each month.

1. Modify the model to take this into account (use the parameters and initial populations shown in Figure 4.23).

2. What effect will this have in the long term? Would you say the system is sensitive to the parameter m?

3. How many rabbits could the hunters kill each month and still have the populations survive in the long-term?

4.5.4 Consider a forest that contains foxes and wolves that compete for the same food resources. If F_n and W_n represent the populations of foxes and wolves, respectively, at the end of month n, a model for their populations is

$$F_{n+1} = 1.2\, F_n - 0.001\, F_n W_n$$
$$W_{n+1} = 1.3\, W_n - 0.002\, F_n W_n.$$

1. Explain the meaning of the parameters 1.2 and 1.3.

2. Why are the parameters -0.001 and -0.002 both negative?

3. Find the equilibria for this system and graphically determine if they are stable or unstable (consider only values up to $n = 25$).

4. Is this model sensitive to the initial populations? Why or why not?

4.6   Epidemics

Consider a community of 1000 persons in which three members get sick with the flu. The following week, five new cases of the flu are reported. We are interested in modeling the spread of the disease through the community.

Consider the following assumptions:

1. No one enters or leaves the community and no one in the community has contact with anyone outside the community.

2. Each person is either susceptible, S (able to get the flu); infected, I (currently has the flu and able to spread it); or removed, R (already had the flu and is not able to get it again). Initially every person is either susceptible or infected.

3. A susceptible person can get the flu only by contact with an infected person.

4. Once a person gets the flu, he or she cannot get it again.

5. The average duration of the flu is 2 weeks, during which time an infected person can spread the disease to a susceptible person.

The model we are going to build is called an *SIR* model (Allman, 2004). We begin by dividing the population into three categories—susceptible, infected, and removed—as described in assumption (2). People move between these three categories as illustrated in Figure 4.28.

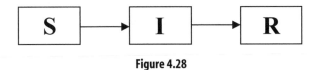

Figure 4.28

Let S_n, I_n, and R_n represent the numbers of persons who are susceptible, infected, and removed, respectively, at the end of week n. As in previous models, we begin by modeling the *change* of these variables using differential equations.

Let's begin by modeling R_n. Assumption (5) says that the average duration of the flu is 2 weeks. This means that, on average, about half the infected persons will be healed (or removed) each week. Therefore, if we let $\gamma = 0.5$, a difference equation for R_n is

$$\Delta R_n = R_{n+1} - R_n = \gamma I_n \tag{4.18}$$

yielding the model

$$R_{n+1} = R_n + \gamma I_n. \tag{4.19}$$

The parameter γ is called the *removal rate* and represents the proportion of the infected persons who are removed each week. The quantity γI_n can be thought of as the number of "newly removed" persons each week.

Now for I_n. This quantity will increase due to some persons who are newly infected as a result of the interactions of the susceptible and infected persons, and it will decrease due to the newly removed persons. Therefore, a difference equation is

$$\Delta I_n = I_{n+1} - I_n = \alpha S_n I_n - \gamma I_n. \tag{4.20}$$

The first term, $\alpha S_n I_n$, models the number of interactions of the susceptible and infected persons, similar to the nonlinear predator–prey model. This quantity can be thought of as the number of "newly infected" persons. Notice that the number of newly infected persons in a week is a product of the number of infected and susceptible persons in the previous *week*. The parameter α is called the *transmission coefficient*. It is a measure of the likelihood that an interaction between an infected person and a susceptible person will result in an infection and is most likely a very small number. Equation (4.20) yields the model

$$I_{n+1} = I_n + \alpha S_n I_n - \gamma I_n. \tag{4.21}$$

Lastly, to model S_n, note that this quantity is decreased only by the number of newly infected persons; it does not increase. Thus a difference equation is

$$\Delta S_n = S_{n+1} - S_n = -\alpha S_n I_n. \tag{4.22}$$

Therefore the model is

$$S_{n+1} = S_n - \alpha S_n I_n. \tag{4.23}$$

The overall model is given by the three equations

$$S_{n+1} = S_n - \alpha S_n I_n$$
$$I_{n+1} = I_n + \alpha S_n I_n - \gamma I_n$$
$$R_{n+1} = R_n + \gamma I_n.$$

Now we determine the values of the parameters α and γ. We already noted that γ is the proportion of infected persons removed each week. In this case, $\gamma = 0.5$. In general,

$$\gamma = \frac{1}{\text{average duration of the infectious period}}.$$

To find the value of α, we need to use the fact that we started with three sick persons and had five new cases the first week. In terms of the variables, this means that in week 0,

$$I_0 = 3 \text{ and } S_0 = 997.$$

In week 1, the number of "newly infected" persons is five, so

$$5 = \alpha S_0 I_0 = \alpha(3 \cdot 997) \tag{4.24}$$

$$\Rightarrow \alpha = \frac{5}{3 \cdot 997} = 0.00167. \tag{4.25}$$

Now we can implement the model.

1. Rename a blank worksheet "**Epidemics**" and format it as in Figure 4.29. Copy row 8 down to row 32 to model 25 weeks.

	A	B	C	D	E	F
1		**Population**	1000			
2		**Transmission Coefficient**	0.00167			
3		**Removal Rate**	0.5			
4						
5					**Newly**	**Newly**
6	**Week**	**Susceptibles**	**Infectives**	**Removals**	**Infected**	**Removed**
7	0	=C1-C7	3	0	=$C$2*B7*C7	=$C$3*C7
8	=A7+1	=B7-E7	=C7+E7-F7	=D7+F7	=$C$2*B8*C8	=$C$3*C8

Figure 4.29

2. Next, use the data in columns **Susceptibles**, **Infectives**, and **Removals** to form a graph as in Figure 4.30.

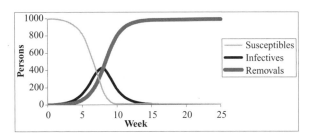

Figure 4.30

Notice that the graph shows that the worst of the epidemic will occur during week 8 when a total of approximately 427 persons will have the flu. Also note that the numerical results show that approximately nine persons will never get the flu. In mathematical notation,

$$\lim_{n \to \infty} S_n \approx 9.$$

The fact that this limit is not zero is typical for an SIR model.

Now let's examine sensitivity to the initial number of cases. It is often the case that at the beginning of an epidemic, the number of initial cases (I_0) is underreported. Using our model we can easily assess this impact.

We will assume that the number of new cases in week 1, five, is accurate. If the number of initial cases changes, the transmission coefficient will also change. Looking back at equation (4.25), we see that

$$\alpha = \frac{\text{number of new cases in week 1}}{(\text{initial number of cases})(\text{population} - \text{initial number of cases})}. \tag{4.26}$$

	B	C
1	Population	1000
2	Transmission Coefficient	=C4/(C7*B7)
3	Removal Rate	0.5
4	Number of New Cases in Week 1	5

Figure 4.31

To implement equation (4.26) modify the worksheet as in Figure 4.31.

Change the number of infectives in week 0 to between 4 and 10 and note how the system changes. Three examples are given in Figure 4.32.

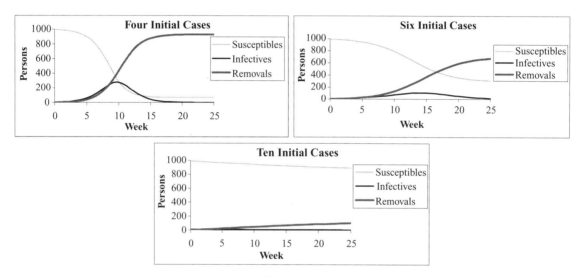

Figure 4.32

Notice that with four initial cases the value of the transmission coefficient decreases, the worst of the epidemic occurs about 2 weeks later, and the severity decreases. At the peak, only 277 persons have the flu and in the long run, about 69 persons never get the flu.

As the number of initial cases increases, the severity decreases. Particularly note that with 10 or more initial cases, the number of infectives never increases. In this case we say that there is no epidemic. We see that if the initial number of infectives is higher than originally thought, the epidemic may not be as bad.

Exercises

4.6.1 In the original model we assumed that the population is constant. Now let's relax this assumption. Suppose that 25 persons move into the community each week starting in week 1. Assume that each of these persons is susceptible.

1. Modify the worksheet to incorporate this influx of persons and describe what happens to the spread of the flu over a 100-week period (use $\alpha = 0.00167$, $\gamma = 0.5$, and $I_0 = 3$).

2. What happens if 100 new persons move in each week?

4.6.2 Suppose the population is a constant 1000, initially 50 persons have the flu ($I_0 = 50$), $\alpha = 0.00167$, and $\gamma = 0.5$. To try to decrease the severity of the epidemic, the community quarantines some of those with the flu. One way to model quarantining is to simply modify the transmission coefficient. For instance, if 25% of those with the flu are quarantined, then only 75% of the interactions between those infected and those susceptible are capable of producing an infected person. Therefore, the number of newly infected persons is given by

$$(0.75\alpha)\, S_n I_n.$$

Thus the new "effective" transmission coefficient is 0.75α.

1. Add a cell to the worksheet **Epidemics** to hold the new parameter "Proportion Quarantined," set it equal to 0.25, and modify the model to incorporate this parameter. Describe what effect quarantining 25% has over not quarantining anyone.

2. Add a scroll bar that allows the user to vary the proportion quarantined between zero and one. Describe what happens as the proportion quarantined changes. What happens if it equals zero? What if it equals one?

3. Add a scroll bar that allows the user to vary the number of initial cases, I_0, between zero and 1000.

4. Find a value of Proportion Quarantined that prevents an epidemic from occurring (meaning the number of infectives never increases) regardless of the value of I_0. Estimate the smallest such value of Proportion Quarantined.

4.6.3 Consider a disease such as the common cold where a person is *not* immune once he or she is "healed." Once healed, a person becomes susceptible again. Such a disease could be modeled with an SIS model as illustrated in Figure 4.33.

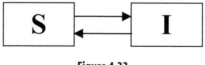

Figure 4.33

1. Devise a set of equations for an SIS model. (**Hint:** There is no "Removed" category. Infected persons are "healed" and then immediately added to the susceptible category.)

2. Implement your model in an Excel worksheet to describe the spread of the common cold through a population of 1000 where initially four persons have the cold. Assume that the cold lasts an average of 2 weeks (use $\alpha = 0.00167$). What do you observe?

For Further Reading

- For an extremely well-written treatise on discrete dynamical systems, see J. T. Sandefur. 1990. *Discrete dynamical systems theory and applications.* New York, NY: Clarendon Press.

- For additional examples of discrete dynamical systems, see M. M. Meerschaert. 1999. *Mathematical modeling.* 2nd ed. San Diego, CA: Academic Press, 141–152.

- For examples of discrete dynamical systems applied to compound interest and mortgage payments, see K. Tung. 2007. *Topics in mathematical modeling.* Princeton, NJ: Princeton University Press, 54–67.

- For additional information on epidemic models, see E. S. Allman and J. A. Rhodes. 2004. *Mathematical models in biology: An introduction.* Cambridge, UK: Cambridge University Press.

- For information on analyzing discrete dynamical systems with eigenvalue and eigenvector methods, see D. C. Lay. 2003. *Linear algebra and its applications.* 3rd ed. Boston, MA: Addison Wesley, 342–353.

- For a much different approach to modeling dynamical systems, see B. Hannon and R. Matthias. 2001. *Dynamic modeling.* New York, NY: Springer.

References

Allman, E. S., and J. A. Rhodes. 2004. *Mathematical models in biology: An introduction.* Cambridge, UK: Cambridge University Press.

Lay, D. C. 2003. *Linear algebra and its applications.* 3rd ed. Boston, MA: Addison Wesley, 342.

Meerschaert, M. M. 1999. *Mathematical modeling.* 2nd ed. San Diego, CA: Academic Press, 127.

CHAPTER 5

Differential Equations

Chapter Objectives

- Use Euler's method to find numerical solutions to differential equations
- Numerically solve systems of differential equations
- Model population growth and military combat with differential equations
- Use eigenvalues to determine behavior of systems of differential equations

5.1   Introduction

Oftentimes it is very easy to describe how fast a quantity changes. For instance, in the bacteria population example in Chapter 4, we observed from the data that the rate at which the culture grows decreases as the population nears 621. This observation led to the difference equation

$$\Delta a_n = a_{n+1} - a_n = b(621 - a_n)\, a_n. \tag{5.1}$$

Note in this example we worked in discrete units of time. In reality, time is continuous, so using discrete time units is a simplification. It is a convenient simplification because a difference equation such as (5.1) is very easy to solve for a_{n+1} in terms of a_n giving a recursive solution.

When using continuous time, we describe change with a *differential equation*. Differential equations are formed in the same basic way as difference equations, but finding their solutions is much more complicated.

To illustrate how differential equations are formed, consider the following observation:

> When a hot cup of coffee is set on a desk, it initially cools very quickly. As the coffee gets closer to room temperature, it cools less quickly.

This simple observation is an example of Newton's law of cooling:

> The rate at which a hot object cools (or a cold object warms) is proportional to the difference between the temperature of the object and the temperature of its surrounding medium.

This law can be translated into the following differential equation:

$$\frac{dy}{dt} = k(y - T)$$

where

$$
\begin{aligned}
y(t) &= \text{ temperature of the object at time } t \\
T &= \text{ temperature of the medium (assumed to be constant)} \\
k &= \text{ constant of proportionality.}
\end{aligned}
$$

This differential equation can be solved using basic techniques, yielding the general solution

$$y(t) = T + Ce^{kt}$$

where C is an arbitrary constant.

Example 5.1.1   Newton's Law of Cooling

Consider a cup of coffee that is initially $100°F$, cools to $90°F$ in 10 min, and sits in a room whose temperature is a constant $T = 60°F$.

The general solution to Newton's law of cooling is $y(t) = T + Ce^{kt}$. To find the specific solution in this case we need to find the values of the constants C and k. The condition $y(0) = 100$ gives

$$100 = 60 + Ce^{k(0)} \quad \Rightarrow \quad C = 40.$$

The condition $y(10) = 90$ gives

$$90 = 60 + 40e^{k(10)} \quad \Rightarrow \quad k \approx -0.02877.$$

Thus the model is

$$y(t) = 60 + 40e^{-0.02877t}.$$

In this chapter, we typically do not try to solve differential equations. Instead, we use a technique called *Euler's method* to numerically approximate solutions and then use the approximation to graphically analyze solutions.

5.2   Euler's Method

Euler's method is a technique for approximating solutions to differential equations. To illustrate the method, consider an autonomous differential equation of the form

$$\frac{dy}{dt} = F(y).$$

Let h be some small positive quantity. Suppose that at time t_0, $y(t_0) = y_0$; at time $t_1 = t_0 + h$, the true value of y is $y(t_1)$.

In the triangle in Figure 5.1, the base has length h and the hypotenuse is on a line with slope $F(y_0)$. Therefore, the height is

$$\text{Height} = hF(y_0).$$

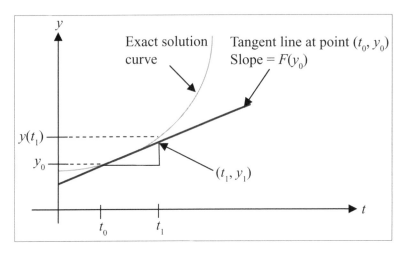

Figure 5.1

The y-coordinate of the base of the triangle is y_0. Thus the y-coordinate of the top of the triangle is

$$y_1 = y_0 + hF(y_0). \tag{5.2}$$

This y-coordinate is an approximation of $y(t_1)$. To approximate $y(t_2)$ where $t_2 = t_1 + h$, we can repeat this process, replacing y_0 with y_1. We could continue to repeat this process to approximate $y(t)$ for any value of $t > 0$ as follows:

$$t_1 = t_0 + h \qquad y_1 = y_0 + hF(y_0)$$
$$t_2 = t_1 + h \qquad y_2 = y_1 + hF(y_1)$$
$$\vdots \qquad\qquad \vdots$$
$$t_{n+1} = t_n + h \quad y_{n+1} = y_n + hF(y_n).$$

Example 5.1.2   Approximating Solutions with Euler's Method

Euler's method is easy to implement in Excel. Here we use it to approximate a numeric solution to the Newton's law of cooling problem in Example 5.1.1, compare it to the exact solution, and examine how the value of h affects the approximation.

1. Rename a blank worksheet "**Euler**" and format it as in Figure 5.2. The formula in cell **C5** comes from the third equation from Example 5.1.1. Add a scroll bar with **min** and **max** of **0** and **1000**, respectively, and set the linked cell to **E1**. Copy row 5 down to row 1004 to calculate values at 1000 different time values.

	A	B	C	D
1	h =	=E1/1000		
2				
3	Time	Approximate	Exact	Error
4	0	100	100	=C4-B4
5	=A4+$B$1	=B4+$B$1*(-0.02877*(B4-60))	=60+40*EXP(-0.02877*A5)	=C5-B5

Figure 5.2

2. Use the data in columns **Time** and **Error** to create a graph similar to those in Figure 5.3. Set the y-axis **min** and **max** to **0** and **0.25**, respectively.

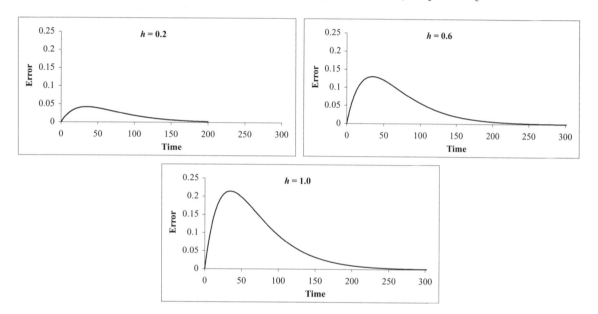

Figure 5.3

Move the slider on the scroll bar left and right to vary the value of h between 0 and 1 and observe what happens to the error. Three examples are shown in Figure 5.3. Note that as h gets smaller, the error decreases; as it gets larger, the error increases. However, note that the error is never greater than 0.25 (the maximum error of approximately 0.214 occurs when $h = 1$ and the exact value is around 75, so the largest percentage error is approximately $0.214/75 \approx 0.29\%$), so our approximate solution is very good.

In general, the smaller h is, the better the approximation. This begs the question, "How do we choose the value of h, especially when we don't know the exact solution?" The simple answer is that it depends on how accurate we want to be and how many calculations we want to make.

For instance, if we want to estimate $y(100,000)$, using $h = 0.01$ would require 10,000,000 iterations of Euler's method. If we want to be very accurate, this may be worth it. If we want a rough approximation, $h = 10$ may be appropriate.

The next two examples illustrate how easy (relatively speaking) it is to model with differential equations.

Example 5.2.2   Logistic Equation

Suppose that 25 panthers are released into a game preserve. Initially the population grows at a rate of approximately 25% per year, but because of limited food supplies, the preserve is believed to support only 200 panthers. We want to model the population over time.

Note that the information given deals with the rate of change. We will create a differential equation to model the rate of change of the population and then use it to approximate the population using Euler's method.

If $y(t)$ represents the population at year t, the initial rate of 25% suggests that we model

$$\frac{dy}{dt} = 0.25y.$$

However, this model does not take into account the fact that the preserve can support only 200 panthers. It seems reasonable to assume that the rate of growth will decrease as y approaches 200. One way to model this is

$$\frac{dy}{dt} = 0.25 \left(1 - \frac{y}{200}\right) y. \tag{5.3}$$

Note that as $y \to 200$, $\left(1 - \frac{y}{200}\right) \to 0$ meaning that $\frac{dy}{dt} \to 0$. Equation (5.3) is called a *logistic differential equation*. Note that this logistic differential equation is very similar to the logistic difference equation we derived for the bacteria population in Chapter 4. The general form of a logistic equation is

$$\frac{dy}{dt} = k \left(1 - \frac{y}{L}\right) y.$$

The parameter L is called the *carrying capacity* and the parameter k is called the *unconstrained* (or *intrinsic*) *growth rate*.

To find a numerical and graphical solution, follow these steps:

1. Rename a blank worksheet "**Logistic**" and format it as in Figure 5.4. Copy row 5 down to row 129 to model 25 years.

	A	B
1	h =	0.2
2		
3	Year	Population
4	0	25
5	=A4+$B$1	=B4+$B$1*(0.25*(1-B4/200)*B4)

Figure 5.4

2. Create a graph as in Figure 5.5.

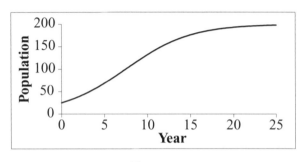

Figure 5.5

The graph shows that the rate of growth slows down as the population approaches 200, as expected. The population reaches the carrying capacity at approximately year 25. Also note that this graph looks very similar to the graph of the bacteria population in Example 4.3.1.

Non-autonomous differential equations (i.e., equations where the right-hand side explicitly depends on both y and t) of the form

$$\frac{dy}{dt} = F(y, t)$$

arise frequently in applications. Euler's method can be easily adapted to approximate solutions to these types of differential equations. The basic algorithm is given by

$$t_{n+1} = t_n + h, \quad y_{n+1} = y_n + hF(y_n, t_n).$$

The next example illustrates an application of a non-autonomous differential equation.

Example 5.2.3   Mixing Solutions

Consider a tank that contains 50 gal of a solution composed of 90% water and 10% alcohol. A second solution containing 50% water and 50% alcohol is added to the tank at the rate of 2 gal per min. At the same time, solution is being drained from the tank at the rate of 3 gal per min. Assuming the tank is continuously stirred, we want to find the amount of alcohol in the tank for times between 0 and 50.

Let $y(t)$ represent the amount of alcohol in the tank at time t. Note that alcohol is being added to the tank and drained out. This suggests a basic model:

$$\frac{dy}{dt} = \text{Rate of change of alcohol (in gal of alcohol/min)}$$
$$= \text{Rate in} - \text{Rate out.}$$

Modeling the rate in is easy. A 50% alcohol solution is being added to the tank at a constant 2 gal per min. Therefore,

$$\text{Rate in} = \left(\frac{0.5 \text{ gal alcohol}}{1 \text{ gal solution}}\right)\left(\frac{2 \text{ gal solution}}{\min}\right) = \frac{1 \text{ gal alcohol}}{\min}.$$

Modeling the rate out is a little more complicated because the proportion of the solution that is alcohol is changing. Note that the volume of solution in the tank is decreasing at the rate of 1 gal/min. Therefore, the volume of solution in the tank at time t is simply $50 - t$. The amount of alcohol in the tank at time t is simply y. Therefore, similar to the model for rate in, we have

$$\text{Rate out} = \left(\frac{y \text{ gal alcohol}}{(50 - t) \text{ gal solution}}\right)\left(\frac{3 \text{ gal solution}}{\min}\right) = \frac{3y}{50 - t} \frac{\text{gal alcohol}}{\min}.$$

Thus the differential equation is

$$\frac{dy}{dt} = 1 - \frac{3y}{50 - t}.$$

To find an approximate solution and view it graphically, follow these steps:

1. Rename a blank worksheet "**Mixture**" and format it as in Figure 5.6. Copy row 5 down to row 254.

	A	B
1	h = 0.2	
2		
3	Time	Amt Alcohol
4	0	5
5	=A4+$B$1	=B4+$B$1*(1-3*B4/(50-A4))

Figure 5.6

2. Create a graph as in Figure 5.7. Notice that the amount of alcohol initially increases and then decreases to 0 as time approaches 50.

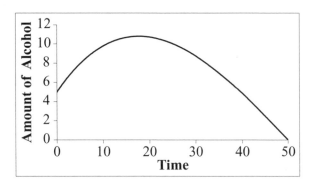

Figure 5.7

Exercises

5.2.1 At time $t = 0$, a yeast culture weighs 0.5 g. Two hours later, it weighs 2 g. The maximum weight of the culture is 8 g.

1. Create a spreadsheet to model the population using a logistic equation. Use a scroll bar to vary the value of k.

2. Use the scroll bar to find a value of k so that the condition $y(2) = 2$ is satisfied.

3. At what time is the weight increasing most rapidly? Support your answer numerically.

5.2.2 A 100-gal tank is full of a solution that contains 15 lb of salt. Starting at time $t = 0$, distilled water is poured into the tank at the rate of 5 gal per min. The solution in the tank is continuously well stirred and is drained out of the tank at the rate of 5 gal per min.

1. Let $y(t)$ represent the amount of salt in the tank at time t. Form a differential equation to model dy/dt.

2. Use Euler's method to estimate the amount of salt in the tank at time $t = 10$.

3. Find the quantity of salt in the solution as $t \to \infty$. Is this what you expect?

5.2.3 In Chapter 2, we modeled the velocity of a free-falling object with air resistance by assuming that the force due to air resistance is proportional to the velocity $(F = k\,v)$. This yielded the differential equation

$$\frac{dv}{dt} + \frac{k}{m}v = g$$

where $g = 9.8$ m/s^2 and $m = $ mass of the object.

1. Suppose $m = 10$ g, $v(0) = 0$, and $k = 3$. Use Euler's method to approximate $\lim\limits_{t \to \infty} v(t)$.

2. Use a scroll bar to examine what happens to $\lim\limits_{t \to \infty} v(t)$ as k changes. Do your results make sense? Why or why not?

3. Add a scroll bar to vary the value of m. What happens to $\lim\limits_{t \to \infty} v(t)$ as m changes? Does this make sense?

5.3 Quadratic Population Model

In this section we model the population of two species with a system of two differential equations. We do not attempt to solve the system; rather, we numerically approximate solutions with Euler's method and analyze the trajectories in the phase plane.

Euler's method for a system of two differential equations is very similar to that in Section 5.2. Consider a two-dimensional system of differential equations with the general form

$$\frac{dx}{dt} = F(x, y), \quad \frac{dy}{dt} = G(x, y).$$

Euler's method for approximating solutions to a system such as this is

$$
\begin{aligned}
t_1 &= t_0 + h & x_1 &= x_0 + hF(x_0, y_0) & y_1 &= y_0 + hG(x_0, y_0) \\
t_2 &= t_1 + h & x_2 &= x_1 + hF(x_1, y_1) & y_2 &= y_1 + hG(x_1, y_1) \\
&\;\;\vdots & &\;\;\vdots & &\;\;\vdots \\
t_{n+1} &= t_n + h & x_{n+1} &= x_n + hF(x_n, y_n) & y_{n+1} &= y_n + hG(x_n, y_n).
\end{aligned}
$$

Example 5.3.1   Competing Foxes and Wolves

Consider a forest that contains foxes and wolves (we will ignore all other animals in the forest). In the absence of any competition, the fox population grows at the rate of 10%

per year and the wolf population at a rate of 25% per year. The forest can support about 10,000 foxes or 6000 wolves. The two species compete for the same resources, but the extent of this competition is not known. We want to know if both species can coexist or whether one will dominate.

To model this scenario, define

$$F(t) = \text{fox population at time } t$$
$$\text{and} \quad W(t) = \text{wolf population at time } t.$$

We can use logistic equations as in Example 5.2.2 to model the two growth rates in terms of the carrying capacities:

$$\frac{dF}{dt} = 0.10 \left(1 - \frac{F}{10{,}000} \right) F$$
$$\frac{dW}{dt} = 0.25 \left(1 - \frac{W}{6000} \right) W.$$

Now to model the effect of competition, we assume that competition decreases the growth rates by an amount proportional to the product of the two populations (this product models the *interaction* of the two species). This yields the system

$$\frac{dF}{dt} = 0.10 \left(1 - \frac{F}{10{,}000} \right) F - c_1 FW$$
$$\frac{dW}{dt} = 0.25 \left(1 - \frac{W}{6000} \right) S - c_2 FW$$

where c_1 and c_2 are some unknown positive parameters. Rewriting this system, we get our model:

$$\frac{dF}{dt} = 0.10F - \frac{0.10}{10{,}000}F^2 - c_1 FW \tag{5.4}$$
$$\frac{dW}{dt} = 0.25W - \frac{0.25}{6000}W^2 - c_2 FW. \tag{5.5}$$

The system described by (5.4) and (5.5) is an example of a *quadratic population model* that has the general form

$$\frac{dx}{dt} = a_1 x + b_1 x^2 + c_1 xy \tag{5.6}$$
$$\frac{dy}{dt} = a_2 y + b_2 y^2 + c_2 xy. \tag{5.7}$$

Let's think about possible values of c_1 and c_2. The xy terms can be thought of as modeling competition *between* species, while the square terms model competition *within* the species. Different species compete for similar, but different resources. Members within a species compete for the exact same resources. Therefore, it seems reasonable that the competition within the species is more intense than the competition between species. Thus $c_i < b_i$. Let's suppose that

$$c_i = \lambda b_i$$

for some $0 < \lambda < 1$.

Example 5.3.2   Solving a Quadratic Population Model

To implement Euler's method to numerically solve the quadratic population model described by (5.4) and (5.5), follow these steps:

1. Rename a blank worksheet "**Quadratic**" and format it as in Figure 5.8. (Note that in this example, species 1 is fox and species 2 is wolf.)

	A	B	C	D
1	**Species 1**		**Species 2**	
2	$a_1 =$ 0.1		$a_2 =$ 0.25	
3	$b_1 =$ = -0.1/10000		$b_2 =$ = -0.25/6000	
4	$c_1 =$ = B5*B3		$c_2 =$ = B5*D3	
5	$\lambda =$ 0.5			

Figure 5.8

2. Add the formulas in Figure 5.9 and copy row 15 down to row 214.

	A	B
13	**Time**	**Species 1**
14	0	1000
15	=A14+$B$12	=B14+$B$12*($B$2*B14+$B$3*B14^2+$B$4*B14*C14)

	C
13	**Species 2**
14	1000
15	=C14+$B$12*($D$2*C14+$D$3*C14^2+$D$4*B14*C14)

Figure 5.9

3. Create a graph as in Figure 5.10. Set the x-axis **min** and **max** to **0** and **10,000**, respectively, and the y-axis **min** and **max** to **0** and **6000**, respectively. The graph shows that the populations start at (1000, 1000) and stop changing at around (9333, 1333). Thus, with these parameters and initial populations, the foxes and wolves can indeed coexist.

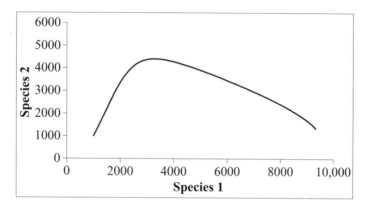

Figure 5.10

4. The initial value of (1000, 1000) is arbitrary, so let's try different values and see what happens. Add the formulas in Figure 5.11 to generate random initial populations. (If these formulas do not work, select **Tools** → **Add-Ins** . . . , select **Analysis ToolPak** and click **OK**.)

	B	C
13	Species 1	Species 2
14	=RANDBETWEEN(0,10000)	=RANDBETWEEN(0,6000)

Figure 5.11

Press the **F9** key several times to try different initial populations. Note that for each one, the populations settle down at around (9333, 1333). Therefore, it appears that the populations can coexist regardless of the initial populations (at least with this value of λ).

The point (9333, 1333) is called an *equilibrium value* of the system. Equilibrium values are central to the analysis of systems of differential equations.

Definition 5.3.1 An *equilibrium value* of a system of differential equations

$$\frac{dx}{dt} = F(x, y), \quad \frac{dy}{dt} = G(x, y)$$

is a point (x_0, y_0) such that $F(x_0, y_0) = 0 = G(x_0, y_0)$.

As with discrete dynamical systems, equilibrium values of systems of differential equations are points on the phase plane that typically attract or repel trajectories. Equilibrium values that attract trajectories (called *stable* or *asymptotically stable*) are important in population models because they indicate what happens to the populations in the long term.

To find equilibrium values of a quadratic population model, we need

$$0 = a_1 x + b_1 x^2 + c_1 xy \tag{5.8}$$
$$0 = a_2 y + b_2 y^2 + c_2 xy. \tag{5.9}$$

Obviously, $(0, 0)$ is one equilibrium value. If $x \neq 0$ and $y = 0$, then (5.8) gives

$$0 = a_1 x + b_1 x^2 = x(a_1 + b_1 x)$$
$$\Rightarrow \quad 0 = a_1 + b_1 x$$
$$\Rightarrow \quad x = -\frac{a_1}{b_1}.$$

Thus

$$\left(-\frac{a_1}{b_1}, 0\right) \tag{5.10}$$

is another equilibrium value. If $x = 0$ and $y \neq 0$, similar calculations give $(0, -\frac{a_2}{b_2})$ as another equilibrium value. Assuming that $x \neq 0$ and $y \neq 0$ requires us to solve the system of equations

$$0 = a_1 x + b_1 x^2 + c_1 xy$$
$$0 = a_2 y + b_2 y^2 + c_2 xy$$

for x and y. Dividing the first equation by x and the second by y and rewriting yields

$$b_1 x + c_1 y = -a_1$$
$$c_2 x + b_2 y = -a_2.$$

Writing this in matrix form gives

$$\begin{bmatrix} b_1 & c_1 \\ c_2 & b_2 \end{bmatrix} \begin{bmatrix} x \\ y \end{bmatrix} = \begin{bmatrix} -a_1 \\ -a_2 \end{bmatrix}.$$

Now, Cramer's rule gives

$$x = \frac{\begin{vmatrix} -a_1 & c_1 \\ -a_2 & b_2 \end{vmatrix}}{\begin{vmatrix} b_1 & c_1 \\ c_2 & b_2 \end{vmatrix}} = \frac{-a_1 b_2 + a_2 c_1}{b_1 b_2 - c_1 c_2} \quad \text{and} \quad y = \frac{\begin{vmatrix} b_1 & -a_1 \\ c_2 & -a_2 \end{vmatrix}}{\begin{vmatrix} b_1 & c_1 \\ c_2 & b_2 \end{vmatrix}} = \frac{-a_2 b_1 + a_1 c_2}{b_1 b_2 - c_1 c_2}.$$

Thus the fourth equilibrium value is

$$\left(\frac{-a_1 b_2 + a_2 c_1}{b_1 b_2 - c_1 c_2}, \frac{-a_2 b_1 + a_1 c_2}{b_1 b_2 - c_1 c_2} \right) \tag{5.11}$$

as long as $b_1 b_2 - c_1 c_2 \neq 0$. This one is particularly important because both the x- and y-coordinates are (potentially) positive. At this equilibrium, the two species coexist and the populations do not change.

Example 5.3.3   Calculating Equilibrium Values
Formula (5.11) is easy to implement in Excel.

1. In the worksheet **Quadratic**, add the formulas in Figure 5.12. We see that the exact value of the equilibrium is $(9333.\overline{3}, 1333.\overline{3})$. Because all trajectories appear to be attracted to this equilibrium, we have graphical evidence that this equilibrium is *stable*.

	A	B
8		**Equilibrium**
9	**x =**	=(-B2*D3+D2*B4)/(B3*D3-B4*D4)
10	**y =**	=(-D2*B3+B2*D4)/(B3*D3-B4*D4)

Figure 5.12

2. Now let's see what happens to the equilibrium value as λ changes. Add a scroll bar, and set the **min** and **max** to **0** and **1000**, respectively, and the linked cell to **B6**. Add the formula in Figure 5.13.

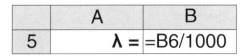

Figure 5.13

Using the scroll bar, we see that for λ between 0 and 0.6, the equilibrium has positive x- and y-coordinates, and it appears to always be stable (we will verify this conclusion analytically in Section 5.6). Thus the two species can coexist. For $\lambda > 0.6$, the y-coordinate of the equilibrium is negative. This means that species 2 (the wolves) will die out and the foxes will dominate. We could consider 0.6 to be the "critical value" of λ.

So, to answer the question, "Can the two species coexist?" we need to determine which is more likely: $\lambda < 0.6$ or $\lambda > 0.6$. It seems reasonable that competition within a species is much greater than competition between species. This means that c_i is much less than b_i, so λ must be very small (i.e., $\lambda < 0.6$). Therefore, we conclude that the two species can coexist.

The general quadratic population model can be adapted to fit a variety of scenarios, as the next example illustrates.

Example 5.3.4   Predator–Prey System

Consider the example of foxes and rabbits discussed in Chapter 4, where rabbits are the sole source of food for foxes. Let's suppose that without rabbits, foxes die at a rate of 8% per month; without foxes, the rabbit population grows at a rate of 4% per month. The presence of rabbits increases the growth of the fox population, and the presence of foxes decreases the growth of the rabbit population.

If $F(t)$ and $R(t)$ represent the populations of foxes and rabbits at time t, respectively, we can model this system using the differential equations (instead of difference equations as in Chapter 4):

$$\frac{dF}{dt} = -0.08F + c_1 FR$$

$$\frac{dR}{dt} = 0.04R + c_2 FR.$$

In Chapter 4, we took $c_1 = 0.0001$ and $c_2 = -0.0003$. Note that this model is a quadratic population model without the square terms (i.e., $b_1 = b_2 = 0$). Entering these values of the parameters into the worksheet **Quadratic**, with an initial population of 100 foxes (species 1) and 1000 rabbits (species 2) and $h = 0.6$, yields a trajectory as shown in Figure 5.14 (you may need to change the scales on the axes to reproduce this graph).

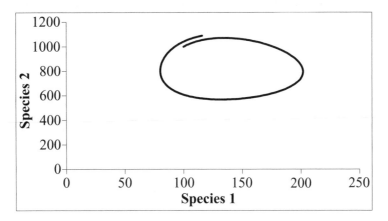

Figure 5.14

Note that the equilibrium is $(133.\overline{3}, 800)$ and the trajectory forms a loop around the equilibrium. (In theory, the trajectory is a closed loop, but because our graph is only an approximation, the loop is open. Using a smaller value of h and more iterations of Euler's method would produce a more accurate graph.) Other initial populations give similar results. Initial populations farther from the equilibrium produce larger loops. This means that the two populations vary more over time. Since the trajectories are not attracted to the equilibrium, we have evidence that it is *unstable*.

These loops mean that the populations oscillate over time. Some biologists argue that this type of model is not realistic because in nature, populations do not tend to oscillate. Rather, they tend to move toward an equilibrium value, as in the fox–wolf model.

Exercises

5.3.1 Investigate sensitivity of the fox–wolf model (5.4) to the carrying capacities. That is, change the carrying capacities a small amount and analyze the resulting model. Does your final conclusion change?

5.3.2 Consider the predator–prey system modeled in Example 5.3.4. Use a scroll bar to investigate the sensitivity of the model to the parameter c_2.

1. What happens to the equilibrium value as c_2 changes?
2. What happens to the variation within each population as c_2 changes?

3. If shelters were built to protect the rabbits from the foxes, what does the model predict might happen to the populations? Would it increase the size of the rabbit population?

5.3.3 The motions of a certain pendulum are described by the system of differential equations:

$$\frac{dx}{dt} = y, \quad \frac{dy}{dt} = -5\sin x - \frac{9}{13}y$$

where $x = \theta$, the angle between the rod and the downward vertical direction, and $y = \frac{d\theta}{dt}$, the speed at which the angle changes. Describe the equilibrium values for this system.

5.3.4 A predator–prey model that takes into account harvesting (i.e., hunting) of the two species is given by

$$\frac{dx}{dt} = a_1 x - b_1 xy - cx$$

$$\frac{dy}{dt} = -a_2 y + b_2 xy - cy$$

where $x(t)$ and $y(t)$ are the populations of the prey and predator species, respectively. All parameters are assumed to be positive. Assuming that $x \neq 0 \neq y$, find a formula for the equilibrium value in terms of the parameters.

5.4   Volterra's Principle

The following scenario is described by Braun (1983, 221) and is reprinted with permission:

> In the mid-1920s the Italian biologist Umberto D'Ancona was studying varia-
> tions in the population of various species of fish that interact with each other.
> In the course of his research, he came across data on percentages-of-total-catch
> of several species of fish that were brought into different Mediterranean ports in
> the years that spanned World War I. In particular, the data gave the percentage-
> of-total-catch of selachians (sharks, skates, rays, etc.), which are not very de-
> sirable as food fish. The data for the port of Fiume, Italy, during the years
> 1914–1923 [are] as follows:
>
> | 1914 | 1915 | 1916 | 1917 | 1918 | 1919 | 1920 | 1921 | 1922 | 1923 |
> |------|------|------|------|------|------|------|------|------|------|
> | 11.9% | 21.4% | 22.1% | 21.2% | 36.4% | 27.3% | 16.0% | 15.9% | 14.8% | 10.7% |
>
> D'Ancona was puzzled by the very large increase in the percentage of selachi-
> ans during the period of the war. Obviously, he reasoned, the increase in the

percentage of selachians was due to the greatly reduced level of fishing during this period, but how does the intensity of fishing affect the fish populations ... It was also a concern to the fishing industry, since it would have obvious implications for the way fishing should be done.

D'Ancona took this problem to the famous Italian mathematician Vito Volterra. Volterra noted that the selachians are predators and the food fish are their prey. So he devised a simple predator–prey model:

$$\frac{dx}{dt} = a_1 x - b_1 xy$$

$$\frac{dy}{dt} = -a_2 y + b_2 xy$$

where $x(t)$ and $y(t)$ are the populations of the prey (the food fish) and the predator (the selachians), respectively, and $a_1, a_2, b_1, b_2 > 0$. To model the impact of fishing, Volterra added another term to each equation:

$$\frac{dx}{dt} = a_1 x - b_1 xy - cx \tag{5.12}$$

$$\frac{dy}{dt} = -a_2 y + b_2 xy - cy. \tag{5.13}$$

The parameter c could be thought of as the proportion of each population caught per unit of time. Rewriting (5.12) and (5.13) we get

$$\frac{dx}{dt} = (a_1 - c)x - b_1 xy$$

$$\frac{dy}{dt} = (-a_2 - c)y + b_2 xy.$$

We see that this model is really just a special case of the quadratic population model studied in Section 5.3. Using the formula derived in Exercise 5.3.4, we see that there is an equilibrium value at

$$\left(\frac{a_2 + c}{b_2}, \frac{a_1 - c}{b_1} \right).$$

Let's examine a system such as this graphically:

1. Rename a blank worksheet "**Volterra**" and format it as in Figure 5.15. Note that the values of these parameters do not come from the data. They are simply arbitrary values we use to illustrate the point we want to make.

	A	B	C	D
1	**Prey**		**Predator**	
2	$a_1 =$ 0.04		$a_2 =$ 0.08	
3	$b_1 =$ 0.0004		$b_2 =$ 0.0001	
4	c = 0.03			
5				
6	**Equilibrium**			
7	x = =(D2+B4)/D3			
8	y = =(B2-B4)/B3			

Figure 5.15

2. Add the formulas in Figure 5.16 to compute numerical solutions using Euler's method. Copy row 13 down to row 1012. Use the calculations to create a graph as in Figure 5.17. As with the parameters, these initial populations are arbitrary.

Notice that we get a loop as in the fox and rabbit predator–prey model. This means that the populations are periodic (like a sine or cosine curve). Mathematically, this means there exists a time T such that

$$x(T) = x(0) \quad \text{and} \quad y(T) = y(0).$$

	A	B
10	h = 0.25	
11	**Time**	**Prey**
12	0	1000
13	=A12+$B$10	=B12+$B$10*($B$2*B12-$B$3*B12*C12-$B$4*B12)

	C
11	**Predator**
12	100
13	=C12+$B$10*(-$D$2*C12+$D$3*B12*C12-$B$4*C12)

Figure 5.16

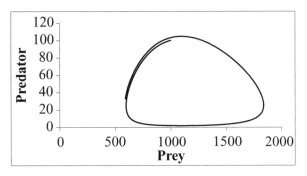

Figure 5.17

3. To approximate the value of T for this model, we use a scroll bar to graph the trajectory in Figure 5.17 over an interval of time so that it makes only one full "loop." Add a scroll bar, and set the linked cell to **I1** and the **min** and **max** to **0** and **1000**, respectively. Add the formula in Figure 5.18.

	F
1	**Time**
2	=I1*B10

Figure 5.18

4. Modify the formulas in Columns **B** and **C** as in Figure 5.19. Copy this modified row 13 down to row 1012. These formulas will calculate the two populations *only* at each time less than the time in cell **F2**. For larger values of time, the formula will return **#N/A**, which means "value not available," and is equivalent to a blank cell.

B
13 =IF(A13<=$F$2,B12+$B$10*($B$2*B12-$B$3*B12*C12-$B$4*B12),NA())

C
13 =IF(A13<=$F$2,C12+$B$10*(-$D$2*C12+$D$3*B12*C12-$B$4*C12),NA())

Figure 5.19

5. Use the scroll bar to find a time so that the graph looks similar to Figure 5.20. Notice that in Figure 5.20, the populations return to where they started, so each has completed one full cycle. The corresponding time is approximately 220. This means $T \approx 220$.

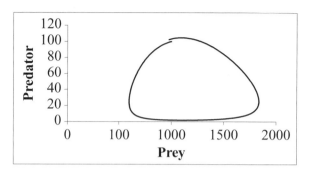

Figure 5.20

6. Next let's look at the average of each population over the period of time $[0, T]$. Add the formulas in Figure 5.21.

	E	F
7	Prey Average =	=SUMIF(B12:B1012,">0")/I1
8	Predator Average =	=SUMIF(C12:C1012,">0")/I1

Figure 5.21

The results of the calculations are shown in Figure 5.22. Comparing the averages to the equilibrium value, we note that they're the same (at least approximately)! If we change the initial populations and the parameters, and each time find T, we observe that the averages are always approximately equal to the equilibrium values. This illustrates Theorem 5.4.1.

	A	B	C	D	E	F
6	Equilibrium					
7	x =	1100			Prey Average =	1104.35
8	y =	25			Predator Average =	25.33

Figure 5.22

Theorem 5.4.1 *Let $(x(t), y(t))$ be a periodic solution of the system described by (5.12) and (5.13) with period T. Define the average values of x and y as*

$$\bar{x} = \frac{1}{T} \int_0^T x(t)dt \ \ and \ \ \bar{y} = \frac{1}{T} \int_0^T y(t)dt.$$

Then

$$\bar{x} = \frac{a_2 + c}{b_2} \ \ and \ \ \bar{y} = \frac{a_1 - c}{b_1}.$$

Theorem 5.4.1 explains D'Ancona's observations. We see that a moderate amount of fishing $(c < a_1)$ increases $\bar{x}$ (the average number of prey, or food fish) and decreases $\bar{y}$ (the average number of predators, or selachians). Conversely, decreased fishing (as happened during World War I) increases the number of selachians and decreases the number of food fish, on average.

The fact that some fishing increases the number of food fish is known as *Volterra's principle*. This principle also applies to other situations, such as applying insecticides as described by Giordano (2003, 435–436), reprinted with permission:

> In 1868 the accidental introduction into the United States of the cottony cushion insect (*Icerya purchasi*) from Australia threatened to destroy the American citrus industry. To counteract this situation, a natural Australian predator, a ladybird beetle (*Novius cardinalis*), was imported. The beetles kept the scale insects down to a relatively low level. When DDT was discovered to kill scale insects, farmers applied it in the hopes of reducing even further the scale insect population. However, DDT turned out to be fatal to the beetle as well and the overall effect of using the insecticide was to increase the numbers of the scale insect.

In this situation, the predator was the beetle and the prey was the cottony cushion insect. Killing both species (as with fishing) decreased the predator and increased the prey. In this case, this end result was unwanted.

Exercises

5.4.1 In the original "fishing" model, we assumed that both the predator and the prey are caught at the same rate. Now consider a refined "insecticide" model where the predator and the prey are killed at different rates where $c_1 \neq c_2$:

$$\frac{dx}{dt} = 0.05x - 0.0004xy - c_1 x$$

$$\frac{dy}{dt} = -0.04y + 0.0001xy - c_2 y.$$

1. Assume that $c_2 = 0.01$. Modify the worksheet **Volterra** to graph trajectories for this refined model.

2. Assume that we start with 100 prey and 50 predator insects. For what values of c_1 is the prey species killed off before the predator species? For what values is the predator species killed off first? For what values is neither species killed off? (**Suggestion:** We could consider the prey

species to be "killed off" first when its population drops below 1 before the population of the predator species does. Graphically, this is when the trajectory touches, or nearly touches, the predator axis (the y-axis). Neither species is killed off when neither population drops below 1.)

5.4.2 Consider the competing foxes and wolves model from Example 5.3.1 with an additional term to model the hunting, or harvesting, of both species:

$$\frac{dF}{dt} = 0.10F - \frac{0.10}{10,000}F^2 - 0.5\left(\frac{0.10}{10,000}\right)FW - HF$$

$$\frac{dW}{dt} = 0.25W - \frac{0.25}{6000}W^2 - 0.5\left(\frac{0.25}{6000}\right)FW - HW.$$

where H is a measure of the amount of hunting.

1. Rewrite this model so it fits the form of a standard quadratic population model given in equation (5.6). Modify the worksheet **Quadratic** to implement this rewritten model. Create a cell for the value of H (suppose we start with 1500 foxes and 1000 wolves).

2. Find the largest value of H so that both coordinates of the equilibrium given by formula (5.11) are positive. What does it mean if both of these coordinates are positive? What if one is negative?

3. Suppose that at time $T = 0$, when people started hunting the two species, the populations were 1500 foxes and 1000 wolves. Suppose at time $T = 50$, the fox population is around 100. Find the value of H necessary for this to happen.

4. What would happen to the two populations if the level of hunting found in part (3) were to continue?

5.5   Lanchester Combat Models

In this section we present an application of systems of differential equations that is very different than the population model presented earlier. However, we will see that the resulting model is not that much different.

During World War I, F. W. Lanchester devised several mathematical models of warfare. Since then, his models have been widely studied and adapted to a variety of scenarios ranging from "isolated battles to entire wars" (Coleman 1983, 109). In this section we derive one such model and analyze it graphically.

Let $A(t)$ represent the number of combatants in army A at time t. The rate at which $A(t)$ changes with respect to time, dA/dt, is affected by several factors including casualties caused by the opposing army, disease, desertions, and reinforcements. For simplicity, we will consider only the first factor.

The rate at which combatants are lost due to casualties caused by the opposing army is often referred to as the *combat loss rate* (CLR). In mathematical notation, this is described by the differential equation

$$\frac{dA}{dt} = -\text{CLR}.$$

Armies are divided into two general categories: conventional and guerilla. A conventional army operates in relatively large units with an identifiable front line, while a guerilla army operates in small units without a front line.

Consider a battle where a conventional army C goes against a smaller guerilla army G. We can describe this scenario with a general system of differential equations:

$$\frac{dC}{dt} = -\text{CLR}_C$$
$$\frac{dG}{dt} = -\text{CLR}_G.$$

Suppose that army C is out in the open in some formation and army G is hidden in the trees of a forest and that each combatant in each army is firing a gun.

Let's consider CLR_C. This is the rate at which combatants in the conventional army are killed or wounded. Obviously the larger $G(t)$ is, the higher the rate. This suggests a proportionality relationship:

$$\frac{dC}{dt} = -g\,G(t). \tag{5.14}$$

The constant of proportionality g is called the *combat effectiveness coefficient* of army G. It can be defined as

$$g = r_G p_G$$

where

$r_G = $ Firing rate of army G (shots/day/combatant)

$p_G = $ Probability that a single shot from army G will hit an opponent.

Now consider CLR_G. It seems reasonable that this is proportional to $C(t)$. However, combatants in army C cannot see those in army G. So they are blindly firing into the forest (this approach has been called "spray and pray"). Thus, the larger $G(t)$ is, the larger the probability that a shot will hit an opponent. This suggests another proportionality relationship:

$$\frac{dG}{dt} = -cC(t)\,G(t). \tag{5.15}$$

The combat effectiveness coefficient c is defined in a similar fashion as g, $c = r_C p_C$. We assume that $r_C = r_G$. Since army C is blindly firing into the forest, the probability that a shot will hit an opponent can be described by

$$p_C = \frac{\text{Area of the exposed part of the body of a single guerilla}}{\text{Total area occupied by the guerillas}}$$

$$= \frac{\text{Area of the exposed part of the body of a single guerilla}}{(\text{Area occupied by a single guerilla}) \cdot G_0}$$

where $G_0 = G(0)$ (the initial number of guerilla combatants). Putting (5.14) and (5.15) together, we have our model:

$$\frac{dC}{dt} = -gG \tag{5.16}$$

$$\frac{dG}{dt} = -cCG \tag{5.17}$$

(the time variable t has been dropped for simplicity). The type of battle modeled here was common in Vietnam where the conventional American and South Vietnamese army fought the guerilla North Vietnamese and Viet Cong army. For this reason, this model is called the "Vietnam" model (Coleman 1983, 111).

Now we use this model to analyze the question, "What ratio of initial combatants $(C_0/G_0 = n)$ is necessary for army C to win?" We say that army C "wins" when army G runs out of combatants first.

To answer this question, we solve the system (5.16) and (5.17) for G in terms of C. Note that

$$\frac{dC}{dG} = \frac{-gG}{-cCG} = \frac{g}{cC}.$$

Cross multiplying, we get

$$g\,dG = cC\,dC. \tag{5.18}$$

Now, integrating both sides of (5.18) yields

$$gG = \frac{1}{2}cC^2 + M$$

where M is an arbitrary constant. Dividing by g gives

$$G = \frac{c}{2g}C^2 + \frac{M}{g}. \tag{5.19}$$

Equation (5.19) does not give us $G(t)$ or $C(t)$ in terms of t, but it does give us a *relation* between $G(t)$ and $C(t)$. We can use this to answer the question graphically.

Now, to find the value of M we need to use the conditions that $G(0) = G_0$ and $C(0) = C_0$. Evaluating (5.19) at $t = 0$ gives

$$G(0) = \frac{c}{2g}C(0)^2 + \frac{M}{g}$$

$$\Rightarrow \quad G_0 = \frac{c}{2g}C_0^2 + \frac{M}{g}$$

$$\Rightarrow \quad M = gG_0 - \frac{c}{2}C_0^2.$$

To implement this model and analyze it to find the value of n so that army C "wins," follow these steps:

1. Rename a blank worksheet "**Vietnam**" and format it as in Figure 5.23. Copy row 11 down to row 109. Note that the values of C in column **A** are in increments of 15 (i.e., they aren't calculated using any formula). The corresponding values of G in column **B** are calculated using equation (5.19). The values of the parameters shown in the figure are those suggested by Coleman (1983, 118).

2. Add a scroll bar, and set the linked cell to **D1** and the **min** and **max** to **0** and **1500**, respectively. Create a graph similar to those in Figure 5.24.

In 5.24 we see that for $n = 8$ (that is, when army C is initally eight times as large as army G), when $C(t) = 0$, $G(t) \approx 54$. This means that army G "wins." We do not know the time t at which this happens, but this is not important for our analysis.

For $n = 12$, when $G(t) = 0$, $C(t) \approx 1000$. This means that army C "wins." For $n = 10$, $C = 0 = G$ at approximately the same time (in other words, they are both totally destroyed

	A	B
1	Firing Rate =	10
2	Exposed Area =	2
3	Area/Guerilla =	1000
4	p_G =	0.1
5	n =	=D1/100
6	g =	=B1*B4
7	c =	=B1*B2/(B3*B10)
8	M =	=B6*B10-B7/2*A10^2
9	Conventional	Guerilla
10	=B5*B10	150
11	=A10-15	=$B$7/(2*$B$6)*A11^2+$B$8/$B$6

Figure 5.23

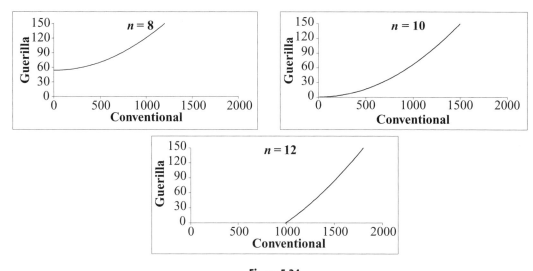

Figure 5.24

at the same time and nobody wins). Using the scroll bar to vary the value of n, we see that army C "wins" for $n > 10$ and "loses" for $n < 10$.

To perform a sensitivity analysis on the model, we change one initial condition or parameter at a time and find the approximate minimum value of n for army C to "win." Changing G_0 or the firing rate, we get the same results as earlier. Table 5.1 shows ranges of values of the other parameters and the associated ranges of the minimum values of n. Summarizing our sensitivity analysis, we conclude that for army C to "win," n must be *at least* 7, and probably more.

	Exposed Area	Area/Guerilla	p_G
Range of Values	1.5–4	500–1750	0.05–0.15
Range of n	11.5–7	7–13.1	7–12.2

Table 5.1

To verify our model we need some data. Table 5.2 (data adapted from a graph in [Coleman 1983, 119]) gives n for several guerilla–conventional conflicts since World War II and the victors. In these data, n is computed using average force strengths over the period of time and does not take into account reinforcement rates or noncombat loss rates. Thus we should be careful about interpreting the data. Nevertheless, we see that the data do tend to support our conclusion.

Conflict	n	Victor
Greece: 1946–1949	9	
Malaya: 1945–1954	18	Conventional
Kenya: 1953	10	
Philippines: 1948–1952	4	
Indochina: 1945–1954	2	
Indonesia: 1945–1947	2	
Cuba: 1958–1959	6	Guerrilla
Laos: 1959–1962	3	
Algeria: 1956–1962	10	
Vietnam: 1959	9	
Vietnam: 1968	6	
Vietnam: 1975	≈ 4	

Table 5.2

Example 5.5.1   Application to the Vietnam War

In the spring of 1968 there were approximately 1,680,000 conventional forces led by the United States and 280,000 guerilla forces led by the North Vietnamese and Viet Cong in Vietnam (Coleman 1983, 120). This means the ratio of conventional forces to the guerilla forces was approximately $\frac{1,680,000}{280,000} = 6$ (i.e., $n \approx 6$). Around this time, General Westmoreland, then commander of U.S. forces in South Vietnam, requested an additional 206,000 troops from President Johnson. Could this have actually helped?

With an additional 206,000 troops, the ratio of conventional forces to the guerilla forces would have increased to $\frac{1,866,000}{280,000} \approx 6.7$ (assuming the size of the guerilla force did not change). This ratio is too small for the conventional forces to win (at least as predicted by our model and supported by the data). Analysis such as this, and a myriad of other factors, caused President Johnson to reject the request for additional troops.

Exercises

5.5.1 Suppose two conventional armies, x and y, are engaged in battle modeled by the system

$$\frac{dx}{dt} = -by$$

$$\frac{dy}{dt} = -cx$$

where $x(t)$ and $y(t)$ represent the number of combatants in armies x and y, respectively. Solving this system yields the relationship

$$y = \sqrt{\frac{c(x^2 - x_0^2)}{b} + y_0^2}.$$

Parameters c and b are the combat effectiveness coefficients of armies x and y, respectively. They represent the strengths of the respective armies. Assume that army y is more powerful than army x (i.e., $b > c$).

1. Assume that $c = \lambda b$ for some $\lambda < 1$. Define $n = \frac{y_0}{x_0}$ (the ratio of initial forces). Create a spreadsheet to graph $y(t)$ versus $x(t)$ for different values of λ and n. (**Suggestions:** Use $x_0 = 100$ with values of x in increments of 1. Also initially take $b = 1.5$. Try different values of b. Does the value of b really make a difference?)

2. Since army y is more powerful, for it to "win" it seems reasonable that y_0 can be less than x_0. How much less can y_0 be so that army y "wins"? To answer this question, define n_0 to be the value of n so that both armies are destroyed (i.e., the graph of $y(t)$ versus $x(t)$ goes through the origin). Choose several values of λ and find the corresponding value of n_0 for each. Find a relationship between λ and n_0. (**Hint:** It's a very simple relationship.)

5.5.2 Suppose two conventional armies, x and y, are engaged in battle using weapons that can be aimed at specific targets (such as rifles) and weapons that can impact a large area (such as grenades). If $x(t)$ and $y(t)$ represent the number of combatants in army x and y, respectively, at time t, a Lanchester model of this battle is

$$\frac{dx}{dt} = -ay - bxy$$

$$\frac{dy}{dt} = -cx - dxy.$$

The parameters a and b represent the effectiveness of army y's specific target weapons and their area weapons, respectively. The parameters c and d have the same meaning for army x.

Army x has a three-to-one numerical superiority at the beginning of the battle. However, army y is better trained and better equipped, and its weapons are more effective. This means that $a > c$ and $b > d$.

1. Assume that $c = 0.1$, $d = 0.001$, $x(0) = 3$, $y(0) = 1$, $a = \lambda c$, and $b = \lambda d$ for some $\lambda > 1$. Let λ_0 be the minimum value of λ necessary for army y to "win" the battle. Use a graphical approach and Euler's method to approximate the value of λ_0.

2. Repeat part (1), but now assume that army x has a 4:1 numerical superiority (i.e., $x(0) = 4$ and $y(0) = 1$). What about a 5:1 superiority?

3. Generalize your results in part (2). Suppose army x has an n:1 numerical superiority where $n \geq 1$. Conjecture a relationship between λ_0 and n.

4. (**Extra Credit**) Prove your conjecture in part (3) analytically (meaning don't use graphs). (**Hint:** Use the differential equations to find $\frac{dy}{dx}$ in terms of a, b, c, d, x, and y. Substitute in the relationships $a = \lambda c$ and $b = \lambda d$. Calculate this slope on the line $x = ny$. Figure out what must be true about λ so that the slope is less than $\frac{1}{n}$.)

5.6   Eigenvalues

An eigenvalue of an $n \times n$ matrix A is a number λ such that the equation

$$A\vec{x} = \lambda\vec{x} \qquad \Leftrightarrow \qquad (A - \lambda I)\vec{x} = \vec{0}$$

has a nontrivial solution (i.e., $\vec{x} \neq \vec{0}$). Eigenvalues play an important role in the analysis of systems of differential equations. In this section, we graphically examine how eigenvalues are related to the behavior of two-dimensional systems of differential equations.

First we consider a linear system of differential equations of the form

$$\frac{dx}{dt} = ax + by$$
$$\frac{dy}{dt} = cx + dy.$$

This system can be written in the matrix form

$$\dot{\vec{x}} = \begin{bmatrix} a & b \\ c & d \end{bmatrix} \vec{x}$$

where $\vec{x} = \begin{bmatrix} x \\ y \end{bmatrix}$ and $\dot{\vec{x}} = \begin{bmatrix} dx/dt \\ dy/dt \end{bmatrix}$.

Let's find a formula for the eigenvalues of the matrix $A = \begin{bmatrix} a & b \\ c & d \end{bmatrix}$ (called the *coefficient matrix*). To calculate eigenvalues, we use the fact that λ is an eigenvalue of A if and only if

$$\det(A - \lambda I) = 0$$

where I is the $n \times n$ identity matrix. So,

$$\det\left(\begin{bmatrix} a & b \\ c & d \end{bmatrix} - \lambda \begin{bmatrix} 1 & 0 \\ 0 & 1 \end{bmatrix} \right) = \det \begin{bmatrix} a - \lambda & b \\ c & d - \lambda \end{bmatrix} = (a - \lambda)(d - \lambda) - cb.$$

Setting this determinate equal to 0 results in the following quadratic equation in λ:

$$\lambda^2 + (-a - d)\lambda + (ad - cb) = 0.$$

Solving this equation using the quadratic formula gives

$$\lambda = \frac{(a + d) \pm \sqrt{(a + d)^2 - 4(ad - cb)}}{2}. \tag{5.20}$$

Notice that there's no guarantee that what's under the square root (called the *determinate*) will be positive, so it's possible that the eigenvalue(s) could be complex numbers.

Example 5.6.1   Calculating Eigenvalues
Consider the system

$$\dot{\vec{x}} = \begin{bmatrix} 1 & -2 \\ 3 & -4 \end{bmatrix} \vec{x}. \tag{5.21}$$

Notice that the origin $(0, 0)$ is the only equilibrium value of this system. The origin will always be an equilibrium value of a linear system such as this.

We will create a spreadsheet to graphically determine whether this equilibrium value is stable or unstable using Euler's method and calculate the eigenvalues using formula (5.20).

1. Rename a blank worksheet "**Eigenvalues**" and format it as in Figure 5.25. Copy row 8 down to row 207. The initial values are random numbers between -1 and $+1$.

	A	B	C
1		**Matrix**	
2	1	-2	
3	3	-4	
4			
5	h =	0.03	
6	t	x	y
7	0	=2*RAND()-1	=2*RAND()-1
8	=A7+$B$5	=B7+$B$5*($A$2*B7+$B$2*C7)	=C7+$B$5*($A$3*B7+$B$3*C7)

Figure 5.25

2. Create a graph as in Figure 5.26 (the large dot shows the initial values). Set the *x*- and *y*-axis **min** and **max** to −1 and +1, respectively. (**Note:** Your graph will probably look considerably different than this because of the random initial values.)

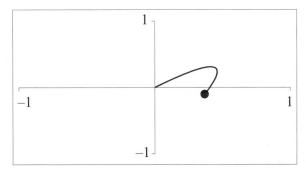

Figure 5.26

Press the **F9** key several times. Each time, new initial values are generated. Notice that the trajectories are always attracted to the origin. In this case we say the origin is *asymptotically stable* or that it is an *attractor*. If the trajectories always move away from the origin, we say the origin is *unstable* or that it is a *repellor*.

3. Add the formulas in Figure 5.27 to compute the eigenvalues. Note that the eigenvalues are −1 and −2.

Using the spreadsheet we can find the eigenvalues and determine the stability of the origin for the systems with the coefficient matrices shown on the next page. For each one, note

	D	E
1	**(a+d)/2 =**	=COMPLEX((A2+B3)/2,0)
2	**(a+d)²-4(ad-cb) =**	=(A2+B3)^2-4*(A2*B3-A3*B2)
3	**Sqrt =**	=IF(E2<0,COMPLEX(0,SQRT(-E2)/2),COMPLEX(SQRT(E2)/2,0))
4		**Eigenvalues**
5	=IMSUM(E1,E3)	=IMSUB(E1,E3)

Figure 5.27

whether the origin is an attractor or a repellor and observe the signs of the real parts of the eigenvalues:

$$\begin{bmatrix} -2 & -5 \\ 1 & 4 \end{bmatrix}, \begin{bmatrix} 7 & -1 \\ 3 & 3 \end{bmatrix}, \begin{bmatrix} -3 & 2 \\ -1 & -1 \end{bmatrix}, \begin{bmatrix} 3 & 1 \\ -2 & 1 \end{bmatrix}, \begin{bmatrix} -3 & -9 \\ 2 & 3 \end{bmatrix}.$$

Note that for the last matrix, the eigenvalues are purely complex numbers (i.e., the real parts are 0). These trajectories form closed loops around the origin. In this case the origin is called a *center*. The graph in the worksheet may not show closed loops due to the fact that Euler's method only gives approximations.

Our observations are summarized in Table 5.3. The main point of these results is that to determine whether the origin is an attractor or a repellor, we need to know only the eigenvalues; a graph of the trajectories is not needed. Note that these results apply to systems of more than two dimensions as well. In the case that we have one eigenvalue that is repeated, things get more complicated. (For a more detailed analytical analysis of these results, see, for example, Boyce and DiPrima 2001, 368.)

Real Part of Eigenvalues	Origin Is Attractor or Repellor
Both negative	Attractor
Both positive	Repellor
One positive, one negative	Repellor
Both 0	Neither (origin is a center)

Table 5.3

Now let's apply these results to a nonlinear system. Let $f : R \to R$ be a function. Using Taylor series, under the right conditions, we may write

$$f(x) = f(x_0) + f'(x_0)(x - x_0) + \frac{f''(x_0)}{2!}(x - x_0)^2 + \frac{f'''(x_0)}{3!}(x - x_0)^3 + \cdots \qquad (5.22)$$

where x_0 is some number at which f and its derivatives are defined. Equation (5.22) gives us a way of computing values of $f(x)$ using series. If f is a complicated function, a series may be easier to work with. The square, cube, and other similar terms are generically called *higher-order terms* (HOTs).

Now suppose we want to evaluate $f(x)$ for some x near x_0. This means that $|x - x_0|$ is a small number (typically less than 1). Thus $(x - x_0)^2$, $(x - x_0)^3$, ... are all even smaller numbers so that HOTs ≈ 0. From (5.22), we get

$$f(x) \approx f(x_0) + f'(x_0)(x - x_0). \tag{5.23}$$

Stated another way, for x near x_0, we can ignore the HOTs. Equation (5.23) gives us a way of approximating $f(x)$ using a very simple linear function. Note that equation (5.23) is simply a different form of equation (5.2), which was used to motivate Euler's method.

Now consider a function of two variables, $F : R^2 \to R^2$, which has the general form

$$F(x, y) = \begin{bmatrix} F_1(x, y) \\ F_2(x, y) \end{bmatrix}. \tag{5.24}$$

Using a two-dimensional version of a Taylor series, if (x_0, y_0) is some point, we may write

$$F(x, y) = F(x_0, y_0) + J(x_0, y_0) \left(\begin{bmatrix} x \\ y \end{bmatrix} - \begin{bmatrix} x_0 \\ y_0 \end{bmatrix} \right) + \text{HOTs}$$

where

$$J(x_0, y_0) = \begin{bmatrix} \frac{\partial F_1}{\partial x} & \frac{\partial F_1}{\partial y} \\ \frac{\partial F_2}{\partial x} & \frac{\partial F_2}{\partial y} \end{bmatrix} \Bigg|_{(x_0, y_0)}.$$

The matrix of partial derivatives,

$$J = \begin{bmatrix} \frac{\partial F_1}{\partial x} & \frac{\partial F_1}{\partial y} \\ \frac{\partial F_2}{\partial x} & \frac{\partial F_2}{\partial y} \end{bmatrix},$$

is called the *Jacobian matrix* (or simply the *Jacobian*). As with the one-dimensional version, if (x, y) is close to (x_0, y_0), then HOTs $\approx \vec{0}$ so that

$$F(x, y) \approx F(x_0, y_0) + J(x_0, y_0) \left(\begin{bmatrix} x \\ y \end{bmatrix} - \begin{bmatrix} x_0 \\ y_0 \end{bmatrix} \right). \tag{5.25}$$

Equation (5.25) gives us a way of approximating values of $F(x, y)$ using a relatively simple linear matrix function.

Example 5.6.2 Application to a Quadratic Population Model

Consider a quadratic population model with the form

$$\begin{bmatrix} \frac{dx}{dt} \\ \frac{dy}{dt} \end{bmatrix} = \begin{bmatrix} a_1 x + b_1 x^2 + c_1 xy \\ a_2 y + b_2 y^2 + c_2 xy \end{bmatrix}. \tag{5.26}$$

The right-hand side of (5.26) is a function of the form in (5.24). The Jacobian for this function is

$$J = \begin{bmatrix} a_1 + 2b_1 x + c_1 y & c_1 x \\ c_2 y & a_2 + 2b_2 y + c_2 x \end{bmatrix}.$$

Now suppose that (x_0, y_0) is an equilibrium value of (5.26). This means that the right-hand side of (5.26) is $\vec{0}$. So, (5.25) gives

$$\begin{bmatrix} \frac{dx}{dt} \\ \frac{dy}{dt} \end{bmatrix} \approx \begin{bmatrix} a_1 + 2b_1 x_0 + c_1 y_0 & c_1 x_0 \\ c_2 y_0 & a_2 + 2b_2 y_0 + c_2 x_0 \end{bmatrix} \left(\begin{bmatrix} x \\ y \end{bmatrix} - \begin{bmatrix} x_0 \\ y_0 \end{bmatrix} \right) \tag{5.27}$$

for (x, y) near (x_0, y_0). Again, note that the right-hand side of (5.27) is linear. This means that the behavior of the system for (x, y) near (x_0, y_0) is approximately linear. Thus we can use the eigenvalues of $J(x_0, y_0)$ and Table 5.3 to analyze the behavior of the system near the equilibrium!. The only exception to this rule is when both eigenvalues have a real part of 0. In this case, the equilibrium value could be a repellor, attractor, or center, so graphical analysis is necessary. The matrix $J(x_0, y_0)$ is sometimes called a *linear approximation* to the system.

Now let's implement this idea to analyze quadratic population models.

1. In the worksheet **Quadratic** from Section 5.3, add the formulas in Figure 5.28 to calculate $J(x_0, y_0)$ where (x_0, y_0) is the equilibrium value calculated in cells **B9** and **B10**.

	C	D
8	Jacobian at Equilibrium	
9	=B2+2*B3*B9+B4*B10	=B4*B9
10	=D4*B10	=D2+2*D3*B10+D4*B9

Figure 5.28

2. Next add the formulas in Figure 5.29 to calculate the eigenvalues of $J(x_0, y_0)$ and determine whether the equilibrium is an attractor.

	E	F
1	(a+d)/2 =	=COMPLEX((C9+D10)/2,0)
2	(-a-d)²-4(ad-cb) =	=(C9+D10)^2-4*(C9*D10-C10*D9)
3	Sqrt =	=IF(F2<0,COMPLEX(0,SQRT(-F2)/2),COMPLEX(SQRT(F2)/2,0))
4		Eigenvalues
5	=IMSUB(F1,F3)	=IMSUB(F1,F3)
6		
7	Attractor?	
8	=IF(AND(IMREAL(E5)<0,IMREAL(F5)<0),"Yes","No")	=-B2/B3

Figure 5.29

In Example 5.3.1 we modeled the populations of foxes and rabbits in a forest with the system

$$\frac{dF}{dt} = 0.10F - \frac{0.10}{10,000}F^2 - c_1 FW$$
$$\frac{dW}{dt} = 0.25W - \frac{0.25}{6000}W^2 - c_2 FW$$

where we assumed that $c_1 = \lambda\frac{0.10}{10,000}$ and $c_2 = \lambda\frac{0.25}{6000}$ for some $\lambda < 1$. Graphically, we concluded that the two species could coexist when $\lambda < 0.6$. We can use these new formulas to confirm this conclusion.

Enter the values of the parameters of this model into the modified worksheet **Quadratic** and use the scroll bar to vary the value of λ. Note that for $\lambda < 0.6$, both coordinates of this equilibrium are positive, and it is an attractor. Thus both species can coexist. For $\lambda > 0.6$, the y-coordinate of the equilibrium is negative, but the equilibrium is a repellor. Graphically it appears that one species dominates while the other dies out. We examine this more in the following exercises.

Exercises

5.6.1 When computing equilibrium values for the quadratic population model, we found that there is another at $\left(-\frac{a_1}{b_1}, 0\right)$.

1. Modify the worksheet **Quadratic** to calculate this equilibrium value and its associated eigenvalues.

2. For the fox–wolf model, what happens to the stability of this equilibrium value as λ changes?

3. What does your answer in part (2) mean in terms of the coexistence of the foxes and wolves for $\lambda > 0.6$?

5.6.2 The models we constructed for competitive species had the form

$$\frac{dx}{dt} = a_1 x - \frac{a_1}{L_1} x^2 - \lambda \frac{a_1}{L_1} xy$$

$$\frac{dy}{dt} = a_2 y - \frac{a_2}{L_2} y^2 - \lambda \frac{a_2}{L_2} xy$$

where a_1 and a_2 are the intrinsic growth rates, L_1 and L_2 are the carrying capacities of the respective species, and $\lambda < 1$. This is a special form of a quadratic population model. We have shown that models such as this have three equilibrium values where at least one coordinate is nonzero (see formulas (5.10) and (5.11)).

1. Suppose $L_1 = L_2$ (i.e., the environment can support the same number of each species). Modify the worksheet **Quadratic** to calculate the values of these three equilibrium values and determine whether each is stable or unstable. Be sure to include a cell for the value of L_1.

2. Now suppose $a_1 = 0.1$ and $a_2 = 0.2$. Experiment with different values of L_1 and observe what happens to the stability of the three different equilibrium values for $0 < \lambda < 3$. (Use any initial populations you want.) How does the value of L_1 affect the behavior?

3. Now switch the values of a_1 and a_2. What do you observe?

4. Generalize your results. If $\lambda < 1$, what happens to the two populations? If $\lambda > 1$, what happens to the two populations? What does this mean about how the relative strength of competition between species compared to competition within a species affects the ability of the species to coexist?

5.6.3 The motions of a certain pendulum are described by the following system of differential equations:

$$\frac{dx}{dt} = y, \quad \frac{dy}{dt} = -5 \sin x - \frac{9}{13} y$$

where $x = \theta$, the angle between the rod and the downward vertical direction, and $y = \frac{d\theta}{dt}$, the speed at which the angle changes. Use eigenvalues to determine the stability for the equilibrium values $(0,0)$, $(\pi,0)$, and $(2\pi,0)$.

For Further Reading

There are more books and articles written on differential equation models than any other type of model. Here are a few suggestions:

- For a good introduction to setting up and solving elementary differential equations, see W. E. Boyce, and R. C. DiPrima, 2001, *Elementary differential equations and boundary value problems*, 7th ed., John Wiley & Sons, New York, NY.

- For another good introduction, see G. Ledder, 2005, *Differential equations: A modeling Approach*, McGraw-Hill, Boston, MA.
- For applications of differential equations to a wide variety of scenarios, see *Modules in applied mathematics, Vol. 1, Differential equation models*, ed. W. F. Lucas, 1983, Springer-Verlag, New York, NY.
- For several classic differential equation models, see R. Haberman, 1998, *Mathematical models—mechanical vibrations, population dynamics, and traffic flow*, Society for Industrial and Applied Mathematics, Philadelphia, PA.
- For more examples of differential equation models, see C. L. Dym, 2004, *Principles of mathematical modeling*, 2nd ed., Elsevier, Boston, MA.

References

Boyce, W. E., and R. C. DiPrima. 2001. *Elementary differential equations and boundary value problems*. 7th ed. New York, NY: John Wiley & Sons, 368.

Braun, M. 1983. Why the percentage of sharks caught in the Mediterranean Sea rose dramatically during World War I. In *Modulus in applied mathematics, Vol. 1, Differential equation models*, ed. W. F. Lucas, 221. New York, NY: Springer-Verlag.

Coleman, C. S. 1983. Combat models. In *Modules in applied mathematics, Vol. 1, Differential equation models*, ed. W. F. Lucas. New York, NY: Springer-Verlag.

Giordano, F. R., M. D. Weir, and W. P. Fox. 2003. *A first course in mathematical modeling*. 3rd ed. Pacific Grove, CA: Thomson Brooks/Cole, a part of Cengage Learning, 435–436. Reproduced by permission. *http://www.cengage.com/permissions*.

CHAPTER 6

Simulation Modeling

Chapter Objectives

- Define and motivate the idea of a simulation model
- Discuss ways of generating pseudorandom numbers
- Use density functions to model random events
- Model various scenarios with simulation models

6.1   Introduction

Mathematical modeling is all about describing systems (a system being a collection of components that operate together). Systems come in two general categories: *deterministic* and *probabilistic*. A deterministic system is one in which the behavior is known once its parameters are known. A probabilistic system is one in which the behavior is determined, in part, by random events. Similarly, models can be put into these same two categories.

An example of a deterministic system is the area under a curve $y = f(x)$ over an interval $[a, b]$. The parameters of this system are the function $f(x)$ and the interval $[a, b]$. Once these parameters are set, the area is determined. Nothing else affects it. We can model this system with a deterministic model using elementary calculus:

$$\text{Area} = \int_a^b f(x)\, dx.$$

Most real-world systems are probabilistic. Inevitably a real-world system involves some type of random event. Probabilistic systems are more difficult to model, so we typically treat them as if they were deterministic and create a deterministic model. Every model of a real-world system we have created to this point in this book has been deterministic.

In this chapter we introduce one very common type of probabilistic model: *simulation*. A simulation, in general, is any model that uses random numbers. Often, simulations are used to imitate some type of real-world behavior, but this does not have to be the case.

There are many reasons why one would construct a simulation model.

1. **System is far too complex to model analytically**. Consider a military air cargo transportation network. This system consists of many components including aircraft of different types, parking spots at airfields, fuel availability at airfields, different types of cargo, and so on. It seems impossible to construct an analytical model that incorporates all of these components. Many simplifications would be needed, resulting in a very low-fidelity model.

2. **It may be difficult, costly, or dangerous to collect data for creating an empirical model**. Consider a hospital emergency room. If we were interested in modeling the waiting time of patients in terms of the number of doctors on staff, we could vary the number of doctors from week to week, collect data on waiting time, and construct an empirical model. This approach would certainly take much time and could result in dead patients. A more practical approach is to simulate the

behavior of the emergency room on a computer and vary the number of doctors in the simulation.

3. **The system may not yet exist**. Consider an aircraft on the drawing board. If we want to study the drag on the fuselage, we cannot go fly it and collect data because it doesn't yet exist! We could simulate its behavior using a computer based on its design.

4. **System may contain random events that we do not want to oversimplify**. Consider a checkout line at a supermarket where customers arrive at an average of two per minute and the cashier can service an average of three customers per minute. If we want to model the waiting time of customers, we might be tempted to say that they won't have to wait at all because the service rate is greater than the arrival rate. This is a vast oversimplification. Instead, we simulate the arrival and service of customers and analyze the waiting time.

In this chapter, we focus on *Monte Carlo simulations*. These simulations get their name from the fact that they are often used to study games of chance (such as those played in Monte Carlo). These simulations consist of three basic steps:

1. Construct a model that uses random numbers.
2. Evaluate, or "run," the model many times (possibly hundreds or thousands) using different random numbers each time.
3. Statistically analyze the results.

One advantage of using simulations to study real-world behavior is that it allows the modeler to test "what-if" scenarios at very little cost. For example, in a simulation of a hospital emergency room, we could easily change the number of doctors or nurses on staff and observe the results. We could simulate many months or years of time in a matter of a few minutes at very little cost and no danger to anyone.

In this chapter we present a wide range of common types of simulation models and discuss some important topics related to the construction of simulation models.

6.2   Basic Examples

In this section we illustrate some of the basic concepts involved with a Monte Carlo simulation with three different examples.

Example 6.2.1   Flipping a Coin

Here we approximate the probability of getting at least 7 tails when a coin is flipped 10 times.

Algorithm

1. Simulate 10 random flips of a coin and determine the number of tails obtained. This is one "trial."
2. Repeat for 200 trials.
3. Calculate $P(\text{getting at least 7 tails}) \approx \dfrac{\text{Number of trials with at least 7 tails}}{\text{Total number of trials}}$.

To implement this algorithm, follow these steps:

1. Rename a blank worksheet "**Coins.**" Format the worksheet as shown in Figure 6.1. The formula in **B2** will select an integer between 0 and 1 with equal probability. An output of 1 indicates a tail and a 0 indicates a head. Copy the formulas in **B2** to the range **B2:K3** and then copy row 3 down to row 201. (**Note:** If Excel returns the #NAME? error in cell **B2**, install and load the **Analysis ToolPak** add-in.)

	A	B	C	D	E	F	G	H	I	J	K	L
1	Trial Number	1	2	3	4	5	6	7	8	9	10	Total Tails
2	1	=RANDBETWEEN(0,1)										=SUM(B2:K2)

Figure 6.1

2. Add the formula in Figure 6.2 to calculate the number of trials with at least 7 tails and the probability.

	N
1	# successes
2	=COUNTIF (L2:L201,">=7")
3	Probability
4	=N2/200

Figure 6.2

Press the **F9** key several times to repeat this experiment. Note that each time you press **F9**, you will probably get a different probability. This is why a simulation like this can only give an approximation of the true *theoretical* probability.

Example 6.2.2   Area Under a Curve

Here we will construct a probabilistic model of a deterministic system. Particularly we will use a simulation to estimate the area under the curve $y = \sqrt{1 - x^2}$ over the interval $[-1, 1]$. This curve forms the top half of a circle with radius 1, so the area under the curve is exactly $\pi/2$. This simulation could also be seen as a way of estimating the value of π.

A graph of the curve is shown in Figure 6.3 along with a rectangle of height $h = 1$ and width $w = 2$ drawn around it.

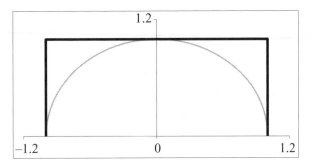

Figure 6.3

In the simulation, we will randomly pick points inside the rectangle and determine whether each one is above or below the curve. We will then estimate the area under the curve using the relationship

$$\frac{\text{Area under the curve}}{\text{Area of the rectangle}} \approx \frac{\text{Number of points under the curve}}{\text{Total number of points}}$$

or equivalently,

$$\text{Area under the curve} \approx \frac{\text{Number of points under the curve}}{\text{Total number of points}}(\text{Area of the rectangle}). \quad (6.1)$$

Algorithm

1. Randomly pick 200 points inside the rectangle.
2. Determine whether each point lies under the curve.
3. Count the number of points under the curve.
4. Use (6.1) to estimate the area under the curve. This is one trial.
5. Repeat for a total of 20 trials.

To implement this algorithm, follow these steps:

1. Rename a blank worksheet "**Area**" and format it as in Figure 6.4. Copy row 6 down to row 205. The formula in cell **C6** tests whether the y-coordinate of the point lies

under the curve $y = \sqrt{1 - x^2}$ that forms the top half of a circle. If it does, then the formula returns a 1. If not, it returns a 0.

	A	B	C
1	a =	-1	
2	b =	1	
3	h =	1	
4			
5	x	y	Under Curve?
6	=$B$1+RAND()*($B$2-$B$1)	=RAND()*$B$3	=IF(B6<SQRT(1-A6^2),1,0)

Figure 6.4

2. Add the formulas in Figure 6.5 to estimate the area. Press the **F9** key several times to repeat the simulation. Note that the estimated area fluctuates quite a bit from repetition to repetition, but the estimates are near the actual area of approximately 1.5708.

	E
1	# Points Under the Curve
2	=SUM(C6:C205)
3	Area Under the Curve
4	=E2/200*(B2-B1)*B3

Figure 6.5

3. To repeat the simulation 20 times, format the spreadsheet as in Figure 6.6 and copy row 4 down to row 22.

	G	H
1	Trial #	Area
2		=E4
3	1	
4	=G3+1	

Figure 6.6

4. Next, highlight the range **G2:H22**, and select **Data → Table. . . .** Select **F1** as the **Column input cell:**, leave the **Row input cell:** blank, and click **OK**. Here's what this does:

(a) The first number in the left column (1 in this case) is "pasted" into the cell **F1**. This causes the random numbers to regenerate, which means another trial is run. The results from the simulation are displayed in the cell **H2** and then copied to the cell next to the 1 (**H3**).

(b) The next number in the first column (2 in this case) is "pasted" into the cell **F1**. The simulation is run again and the results are copied to cell **H4**.

(c) This process is repeated until the bottom of the table is reached.

5. Add the formulas in Figure 6.7 to calculate the average of the 20 trials. Press **F9** several times and note the variation of the average. The average in general is closer to the true area than the single trial case. The average also has less variation.

	E
5	**Average**
6	=AVERAGE(H3:H22)

Figure 6.7

This example illustrates a very important point when analyzing results from simulations:

The more trials, the closer the average value is to the true value.

The point is that if you want to estimate something with a simulation, you will get a more accurate estimate if you use many trials and take an average than if you use only a few trials. This concept is sometimes called the *Law of Large Numbers*.

Example 6.2.3   A Coin-Flipping Game

Here we simulate a simple game of chance and experiment with "what-if" scenarios. (This game is described in F. Hillier and G. Lieberman's *Introduction to Operations Research*, 2001, pp. 1087–1088, and is reproduced by permission of the McGraw-Hill Companies.)

Rules of the Game

1. A single play of the game consists of repeatedly flipping a fair coin until the *difference* between the number of heads tossed and the number of tails is three.

2. You are required to pay $1 for each flip of the coin, and you may not quit during the play of the game.

3. You receive $8 at the end of each play of the game.

Should you play this game? If you play, how much can you expect to win or lose in the long run? The simulation consists of playing the game 500 times and calculating the average winnings. However, the simulation will allow us to change the probability of getting a tail so we can experiment with using a *biased* coin.

1. Rename a blank worksheet "**Coin Flip Game**." Format the worksheet as shown in Figure 6.8. Copy row 10 down to row 57 to simulate flipping the coin up to 50 times. This is one trial. The formulas in column **F** will determine when the game is over. On the flip that the game ends, "Stop" will be displayed. If the game should go on, a blank will be displayed. If the game is already over, "NA" will be displayed.

	A	B	C	D	E
1		Number of Flips =	=COUNTBLANK(F7:F56)+1		
2		Winnings =	=8-C1		
3		Prob of Tail =	0.5		
4					
5		Random		Total	Total
6	Flip	Number	Result	Tails	Heads
7	1	=RAND()	=IF(B7<$C$3,1,0)	=C7	=A7-D7
8	=A7+1	=RAND()	=IF(B8<$C$3,1,0)	=D7+C8	=A8-D8
9	=A8+1	=RAND()	=IF(B9<$C$3,1,0)	=D8+C9	=A9-D9
10	=A9+1	=RAND()	=IF(B10<$C$3,1,0)	=D9+C10	=A10-D1

	F
6	Stop?
7	
8	
9	=IF(ABS(E9-D9)>=3,"Stop","")
10	=IF(F9="",IF(ABS(E10-D10)>=3,"Stop",""),"NA")

Figure 6.8

2. Cells **C1** and **C2** show the results from playing the game once. To store the results from 500 plays of the game, add the formulas in Figure 6.9. Copy row 5 down to row 503 and create a table in the range **H3:J503**. Select **H1** as the column input cell.

	H	I	J
1		Number	
2	Play	of Flips	Winnings
3		=C1	=C2
4	1		
5	=H4+1		

Figure 6.9

3. To calculate the overall average winnings, add the formula in Figure 6.10. Press **F9** several times to repeatedly simulate playing the game 500 times. Note that the average winnings are rarely ever positive and are typically near -1.

	G
1	**Average Winnings**
2	=AVERAGE (J3:J503)

Figure 6.10

4. Next, change the probability of getting a tail to find a value with positive average winnings. Note that for a probability between 0.5 and 0.6, the average winnings are typically negative. However, for a probability of 0.65 or higher, the average is positive. This shows that if a gambler could switch the coin used by the dealer with a biased one with a probability of tails of 0.65 or higher, the gambler would win in the long run.

Exercises

6.2.1 Modify the formulas in the worksheet **Coins** so there are a total of 1000 trials. Press **F9** several times. Compare the variability of the value of the probability in 200 trials as compared to 1000 trials.

6.2.2 Modify the worksheet **Coins** to approximate the probability that the number of tails is between 4 and 7, inclusive. **Hint:** Add a column next to Total Tails to indicate whether the trial is a success (meaning the number of tails is between 4 and 7). The formulas in this column may look similar to that in Figure 6.11. Modify the formula to count the number of successes.

	M
1	**Success?**
2	=IF(AND(L2>=4,L2<=7),1,0)

Figure 6.11

6.2.3 Suppose you roll a fair four-sided die three times. What is the probability that the sum of the three rolls is at least 7? Design a spreadsheet to investigate this question.

6.2.4 Increase the value of h in the worksheet **Area**. How does this affect the quality of the estimation of the area? Also try increasing the number of points selected inside the rectangle. How does this affect the estimate?

6.2.5 In the worksheet **Area**, create a graph of the curve $y = \sqrt{1 - x^2}$ and the rectangle along with a dot for each randomly selected point, similar to Figure 6.12.

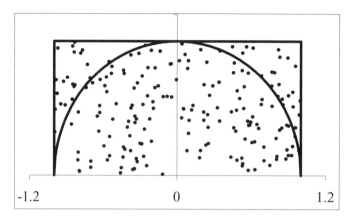

Figure 6.12

6.2.6 A car dealership is sponsoring a contest where the grand prize is a new car. Contestants are to gather tickets that contain the letter "C," "A," or "R" from participating merchants. To win, one must obtain all three letters. If 55% of the tickets contain a "C," 44% contain an "A," and 1% contain an "R," design a simulation to approximate the expected number of tickets you must gather to win the car.

Hints

1. Keep track of the number of each letter you have collected with each ticket.

2. Remember that you win only after you have collected at least one of each letter.

3. Make sure you simulate collecting enough tickets in each trial so that you are almost certain to win in each trial.

4. To calculate the letter on each ticket, consider using formulas as in Figure 6.13.

	B	C
3	**Random**	
4	**Number**	**Letter**
5	=RAND()	=IF(B5<0.55,"C",IF(B5<0.99,"A","R"))

Figure 6.13

6.3   The Birthday Problem

In a class of n students, what's the probability that at least two students will share a birthday (month and day)? This famous problem is known as the *birthday problem*. We assume that birthdays are uniformly distributed throughout the year (i.e., no day is more or less likely to be a birthday than any other day) and we ignore leap years.

Let's start with a class of two students. First, we use complementary events to note that

$$P\left(2 \text{ students share a birthday}\right) = 1 - P\left(2 \text{ students do not share a birthday}\right).$$

We calculate this probability on the right using counting methods. If the two students do not share a birthday, then there are 365 ways to choose the birthday of the first student and 364 ways to choose the birthday of the second. There are $365 \cdot 365$ total number of ways to select the birthdays of two students in general. Thus

$$P\left(2 \text{ students have different birthdays}\right) = \frac{365 \cdot 364}{365 \cdot 365}.$$

Now consider a class of exactly three students. If all three students have different birthdays, then there are 365 ways to choose the first birthday, 364 ways to choose the second, and 363 ways to choose the third. There are $365 \cdot 365 \cdot 365$ ways of choosing the birthdays of three students in general. Thus

$$P\left(3 \text{ students have different birthdays}\right) = \frac{365 \cdot 364 \cdot 363}{365 \cdot 365 \cdot 365}$$

$$= P\left(2 \text{ students have different birthdays}\right) \cdot \frac{363}{365}.$$

We start to see a pattern here:

$$P\left(n \text{ students have different birthdays}\right)$$

$$= P\left(n - 1 \text{ students have different birthdays}\right) \cdot \frac{365 - (n-1)}{365}$$

(assuming $n < 365$). So, we get

$$P\left(\text{at least 2 students share a birthday in a class of } n\right)$$

$$= 1 - P\left(n \text{ students have different birthdays}\right)$$

$$= 1 - P\left(n - 1 \text{ students have different birthdays}\right) \cdot \frac{365 - (n-1)}{365}.$$

This gives a recursively defined solution to the problem that we can easily implement in Excel.

1. Rename a blank worksheet "**Birthday**" and format it as in Figure 6.14. Copy row 3 down to row 101 to calculate the probabilities for a class of up to 100 students.

	A	B	C
1	# students	P(no sharing)	P(sharing)
2	1	1	0
3	=A2+1	=B2*(365-A3+1)/365	=1-B3

Figure 6.14

2. Create a graph of the results as in Figure 6.15. Note that for a class of more than 23 students, the probability is greater than 0.50; for a class of fewer than 23, the probability is less than 0.50.

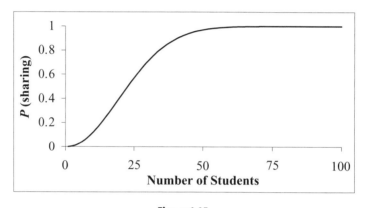

Figure 6.15

We can also use a simulation to estimate a solution to this problem.

Algorithm

1. Randomly generate an integer between 1 and 365 for each student in the class to represent birthdays (1 = January 1, 2 = January 2, and so on).
2. For each day of the year, count the number of students in the class who have that day as their birthday.
3. Determine whether some birthday is shared by at least two students. This is considered a "success."
4. Repeat for 200 trials.
5. Determine the number of successes.
6. Calculate P (at least two people sharing a birthday) $\approx \dfrac{\text{number of successes}}{\text{number of trials}}$.

To implement this algorithm, follow these steps:

1. Rename a blank worksheet "**BirthdaySim**" and format it as in Figure 6.16. Copy row 4 down to row 102 to simulate a class of up to 100 students.

	A	B
1	# **Students=**	23
2	**Person**	**Birthday**
3	1	=IF(A3<=$B$1,RANDBETWEEN(1,365),0)
4	=A3+1	=IF(A4<=$B$1,RANDBETWEEN(1,365),0)

Figure 6.16

2. Add the formulas in Figure 6.17 and copy row 4 down to row 367. These formulas count the number of students who have a birthday on each day of the year and determine whether the trial is a success.

	D	E
1	**Success?**	=IF(COUNTIF(E3:E367,">=2")>=1,1,0)
2	**Day**	**Count**
3	1	=COUNTIF($B$3:$B$102,D3)
4	=D3+1	=COUNTIF($B$3:$B$102,D4)

Figure 6.17

3. Add the formulas in Figure 6.18 to set up a table to store the results of 200 trials and calculate the estimated probability. Copy row 6 down to row 204. Create a table in the range **G4:H204** to store the results from 200 trials. Select **F1** as the column input cell. Press **F9** to repeat the simulation several times. Note that for a class of 23 students the simulation gives a probability of approximately 0.50, as expected.

	G	H
1	**# Successes**	= SUM(H5:H204)
2	**Probability**	= H1/200
3	**Trial**	**Success?**
4		= E1
5	1	
6	=G5+1	

Figure 6.18

One benefit of using a simulation is that we can easily modify it to estimate more complicated probabilities. For instance, if we wanted to estimate the probability that at least 3 students share a birthday in a class of 50 students, we could simply change the value of "# students" to 50 and modify the formula to determine whether the trial is a success as in Figure 6.19.

	D	E
1	**Success?**	=IF(COUNTIF(E3:E367,">=3")>=1,1,0)

Figure 6.19

Exercises

6.3.1 Modify the worksheet **Birthday** to calculate the probability that at least two students in a class of n students share a birthday where there are N "days" in a year. Be sure to include a cell for the value of N.

1. Use the modified spreadsheet to answer this question: In a class of six students, what's the probability that at least two students have a birthday in the same *month*? On what assumption(s) are you basing your solution?

2. Use a table to create a graph of P (at least 2 people share a birthday in a class of 23) versus N for values of N between 1 and 365. What happens to the probability that at least two students share a birthday as N increases? (**Hint:** Set the **Column input cell:** to the cell containing the value of N.)

6.3.2 Modify the worksheet **BirthdaySim** to estimate the solution to a generalization of the birthday problem: If a teacher asks a class of n students to write down an integer between a and b, what's the probability that at least m of them will write down the same number?

1. Your simulation should allow the user to input the values of n, a, b, and m, and automatically calculate the results.

2. Use a total of 500 trials.

3. Assume that $n \leq 200$, $0 \leq a < b \leq 365$, and that the students' choices are uniformly distributed.

4. If the value of m is in cell **E1**, then consider modifying the formula to determine a success as in Figure 6.20.

	D	E
2	**Success?**	=IF(COUNTIF(E4:E368,">="&E1)>=1,1,0)

Figure 6.20

6.4   Random Number Generators

In this section we look at how computers generate "random" numbers. Random numbers are an essential part of all computer simulations. A simple definition of a *list of random numbers* is that it is a list of numbers in which there is *no* pattern. The only way to get a truly random list of numbers is by mechanical means (e.g., numbered balls tumbling in a cage, rolling a die).

A computer generates a list of "random" numbers by using an iterative function where one output becomes the next input. The initial input, called the *seed*, is arbitrary (it is often chosen according to the clock time at which the algorithm begins) and each output becomes a number in the list.

Because the computer uses a deterministic algorithm, there *will* be a pattern to the list of numbers. Therefore, computers can never generate a list of truly random numbers. The "random" numbers they generate are called *pseudorandom* numbers and the algorithm is called a *pseudorandom number generator*. A good list of pseudorandom numbers will, at the very least, have a pattern that is not at all obvious.

Excel has a built-in pseudorandom generator called **RAND** that generates numbers between 0 and 1. First we will create a graph of numbers generated by this function to illustrate what we mean by "no pattern." Rename a blank worksheet "**Rand**," add the formulas in Figure 6.21, and copy row 2 down to row 501 to form a list of 500 pseudorandom numbers.

	A	B
1	**n**	**Random**
2	1	=RAND()

Figure 6.21

Create a graph of Random versus *n* as in Figure 6.22. Press **F9** several times to create new lists of pseudorandom numbers. Note that the graph does not reveal any sort of relationship between the pseudorandom number and its position in the sequence and that no number (or range of numbers) appears more frequently than any other (i.e., there is no pattern). There are many types of statistical tests that can be done to measure the "randomness" of a list of numbers. We do not discuss these here, but our simple graphical analysis indicates that RAND creates a good list of pseudorandom numbers.

Because pseudorandom number generators form the heart of simulations and many security systems, much research has been done in developing and testing generators. Here we present some simple generator algorithms.

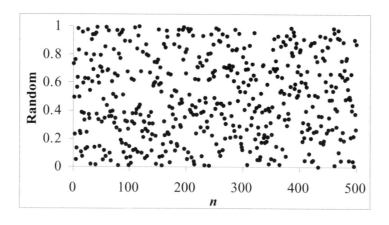

Figure 6.22

Example 6.4.1   Middle-Square Method

This method generates a sequence of n-digit pseudorandom integers. It was first used to simulate neutron collisions in 1946 at Los Alamos Laboratories as part of the Manhattan Project (Giordano et al. 2003, 182). The basic algorithm is as follows:

1. Start with an n-digit number x_0, called the *seed* (n is typically even).
2. Square x_0 to obtain an $2n$-digit number (add leading zero(s) if necessary).
3. Take the middle n digits as the next random number.

To implement the middle-square method to generate a list of 8-digit numbers, rename a blank worksheet "**Mid-Square**" and format it as in Figure 6.23, and copy row 3 down to row 22 to form a list of 20 numbers.

	A	B	C	D	E
1	n	$(x_{n-1})^2$	Chop first two digits	Chop last two digits	x_n
2	0				2660
3	=A2+1	=E2^2	=B3-INT(B3/1000000)*1000000	=INT(C3/100)	=D3

Figure 6.23

With this seed, $x_0 = 2660$, the list appears to be random. However, if $x_0 = 1094$, the sequence begins to repeat after the twelfth number. If $x_0 = 1068$, every number after the twelfth is 0. These illustrate fundamental problems with this method, so it is not widely used.

Example 6.4.2   Linear Congruence

This method uses modular arithmetic to generate pseudorandom integers. Three integers, a, b, and m, are chosen along with a seed x_0. Random integers are then generated using the following function:

$$x_{n+1} = (a \cdot x_n + b) \bmod (m).$$

This method will produce a sequence of up to m integers between 0 and $m-1$, inclusively, before repeating, or *cycling*. For this reason, m is generally a very large integer such as 2^{32} or 2^{64}.

To implement this method using the values of $a = 1$, $b = 3$, $x_0 = 9$, and $m = 8$, rename a blank worksheet "**Linear**," and add the formulas in Figure 6.24. Copy row 6 down to row 25 to create a list of 20 pseudorandom numbers.

	A	B
1	a =	1
2	b =	3
3	m =	8
4	n	x_n
5	0	7
6	=A5+1	=MOD($B$1*B5+$B$2,$B$3)

Figure 6.24

Figure 6.25 shows graphs of the lists of pseudorandom numbers for $a = 1$ and $a = 3$. Note how the lists start to repeat after $n = 8$ and $n = 4$, respectively. This is *not* what we want in a good list of pseudorandom numbers.

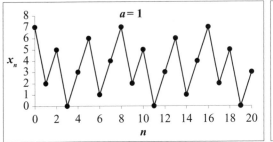 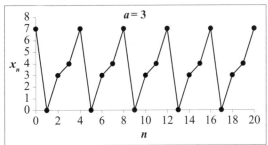

Figure 6.25

Example 6.4.3   Old Excel Algorithm
This algorithm was used by the RAND() function in older versions of Excel and generates a list of up to 1 million different numbers (Microsoft Help and Support Webpage):

1. $x_0 =$ an arbitrary number between 0 and 1

2. $x_{n+1} =$ fractional part of $(9821 * x_n + 0.211327)$

To implement this algorithm with $x_0 = 0.5$, rename a blank worksheet "**Old Rand**," add the formulas in Figure 6.26, and copy row 3 down to row 502.

	A	B
1	n	x_n
2	0	0.5
3	=A2+1	=MOD(9821*B2+0.244327,1)

Figure 6.26

A graph of x_n versus n is shown in Figure 6.27. There does not appear to be any pattern. Other values of x_0 give similar results.

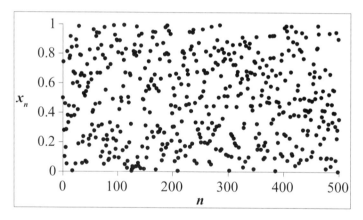

Figure 6.27

Example 6.4.4   New Excel Algorithm

The old Excel algorithm was sufficient for "casual" users (i.e., those who needed fewer than 1 million pseudorandom numbers). However, it did not pass a standard battery of tests for randomness named Diehard, so it was not sufficient for "power" users. A new algorithm was developed that produces a list of up to 10^{13} numbers before cycling. It is a variation on the Linear Congruence method and is based on the idea "that if you take three random numbers on $[0, 1]$ and sum them, the fractional part of the sum is itself a random number on $[0, 1]$" (Microsoft Help and Support Webpage).

The algorithm to generate a list of pseudorandom numbers between 0 and 1, $\{x_n : n = 0, 1, \dots\}$, is as follows:

1. Set a_0, b_0, and c_0 to integer values between 1 and 30,000.
2. x_n = fractional part of $(a_n/30{,}269 + b_n/30{,}307 + c_n/30{,}323)$
3. $a_{n+1} = (171 \times a_n) \bmod (30{,}269)$
4. $b_{n+1} = (172 \times b_n) \bmod (30{,}307)$
5. $c_{n+1} = (170 \times c_n) \bmod (30{,}323)$

To implement this algorithm, rename a blank worksheet "**New Rand**," add the formulas in Figure 6.28, and copy row 3 down to row 502.

	A	B	C	D
1	n	a_n	b_n	c_n
2	0	15843	16235	9842
3	=A2+1	=MOD(171*B2,30269)	=MOD(172*C2,30307)	=MOD(170*D2,30323)

	E
1	x_n
2	=MOD(B2/30269+C2/30307+D2/30323,1)
3	=MOD(B3/30269+C3/30307+D3/30323,1)

Figure 6.28

f A graph of x_n versus n is shown in Figure 6.29. Again note that there does not appear to be any pattern.

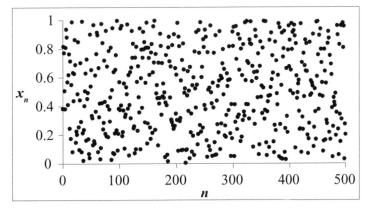

Figure 6.29

Exercises

6.4.1 Suppose we want to generate a list of pseudorandom numbers $\{x_n : n = 1, \ldots , 1000\}$ that have values of a, b, or c where a occurs $p_1 \times 100\%$ of the time, b occurs $p_2 \times 100\%$ of the time, and c occurs $p_3 \times 100\%$ of the time where $p_1 + p_2 + p_3 = 1$.

1. Design a spreadsheet that generates this list $\{x_n\}$. You may use the **RAND** function. Make sure the user is able to input the values of a, b, c, p_1, p_2, and p_3.
2. Use the **COUNTIF** function to verify that the list contains the proper percentage of each number a, b, and c.

6.4.2 Suppose a pseudorandom number generator gives integers between 0 and 9. One way to test whether this generator gives integers with equal frequency is to apply a χ^2 *goodness-of-fit test*. This can be done by generating a long list of integers (say 500) and then counting the number of times each integer appears in the list. These are called the *observed frequencies*, denoted by O. We then calculate the *expected frequencies*, denoted by E, which is the number of times we expect each integer to appear in the list if the integers do indeed occur with equal frequencies. If we have 10 different integers in a list of 500, we would expect each one to appear 50 times.

Then we calculate the "test statistic" χ^2 as follows:

$$\chi^2 = \sum \frac{(O - E)^2}{E}.$$

If this test statistic is "small" (for 10 different integers, "small" is less than 16.9), then we can be 95% confident that the generator gives integers with equal frequencies. If it is "large," then we reject the claim that it gives integers with equal frequencies. (For a more detailed description of this test, see any introductory statistics textbook.)

1. The Excel function **RANDBETWEEN** gives pseudorandom integers between two specified values. Use it to generate a list of 500 integers between 0 and 9. Calculate the χ^2 test statistic as described above and press **F9** several times to get several different lists of integers. Does the **RANDBETWEEN** function appear to give integers with equal frequencies?
2. Use the linear congruence algorithm with $a = 6$, $b = 9$, $m = 10$, and $x_0 = 2$ to generate a list of 500 integers between 0 and 9. Calculate the χ^2 test statistic. Does this linear congruence algorithm appear to give integers with equal frequencies? Try different values of a, b, and x_0. What do you observe?

6.5   Modeling Random Variables

Simulation is a useful tool for modeling the interaction of *random events*. A random event is an activity where we do not know the outcome until it occurs. Therefore, constructing a simulation involves modeling random events. One of the most important concepts used in modeling random events is the *random variable*.

Definition 6.5.1 A *random variable* is a rule for assigning real numbers to the observations of a random event. A *discrete random variable* can take only certain distinct values (such as integers). A *continuous random variable* can take any value within some interval.

In most cases, defining the random variable involved is rather obvious. For instance, suppose we roll a standard six-sided die, and we define the random variable X as the number on the top face of the die. This is an example of a discrete random variable. In another example, suppose we observe customers arriving at a checkout line at the grocery store. We define the random variable Y as the time between customer arrivals (called the inter-arrival times). This is an example of a continuous random variable.

In a simulation of a dice game, we need to generate values of the roll of the dice. In a simulation of a checkout line, we need to generate values of the inter-arrival times. In other words, simulations involve generating values of random variables. In this section we discuss how to do this using the RAND function.

We focus on continuous random variables. One of the most important tools used to model continuous random variables is the *density function*. Any function $f(x)$ is a density function if it satisfies the following two properties:

1. $f(x) \geq 0$ for all $x \in R$

2. $\int\limits_{-\infty}^{\infty} f(x)\,dx = 1$

For a given random variable X, its density function $f(x)$ is used to calculate probabilities regarding the value of X by

$$P(X \leq a) = \int\limits_{-\infty}^{a} f(t)\,dt. \tag{6.2}$$

This means that the probabilities are related to the area under the graph of the density function. Related to the density function is the *cumulative distribution function* (CDF), $F(x)$, defined by

$$F(x) = \int\limits_{-\infty}^{x} f(t)\,dt.$$

From equation (6.2) we see that $F(x) = P(X \leq x)$. In terms of the graph of $f(x)$, $F(x)$ is the area under the curve $y = f(t)$ to the left of x.

Many functions could be density functions. We single out three important types of density functions that occur frequently in applications.

Example 6.5.1   Uniform Density Function

If a random variable X, $a \leq X \leq b$, has a uniform density function (we say X is *uniformly distributed*), the values are "spread out evenly." The RAND function gives values of a pseudorandom variable that is uniformly distributed between 0 and 1. The density function is given by

$$f(x) = \begin{cases} \frac{1}{b-a} & \text{if } a \leq x \leq b \\ 0 & \text{otherwise.} \end{cases}$$

A graph of this density function is shown in Figure 6.30.

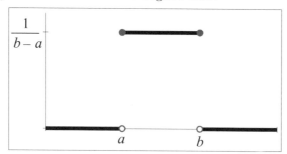

Figure 6.30

The mean (average) of a random variable X with a uniform distribution is $1/2\,(b-a)$ and the standard deviation is $\sqrt{1/12}\,(b-a)$.

Example 6.5.2   Normal Density Function

A normally distributed random variable X has a density function given by

$$f(x) = \frac{1}{\sqrt{2\pi}\sigma} e^{-\frac{(x-\mu)^2}{2\sigma^2}}$$

where μ = mean and σ = standard deviation. The graph of this density function is the familiar "bell curve" shown in Figure 6.31. In the graph we see that the variable X has a higher probability of taking values near the mean μ than farther away.

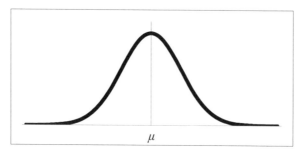

Figure 6.31

Example 6.5.3   Exponential Density Function

An exponential density function is often used to model waiting time between events. It has the form

$$f(x) = \begin{cases} \lambda e^{-\lambda x} & \text{if } x \geq 0 \\ 0 & \text{otherwise.} \end{cases}$$

A graph of this density function is shown in Figure 6.32. In the graph we see that the variable X has a higher probability of taking values near 0 than those much larger than 0.

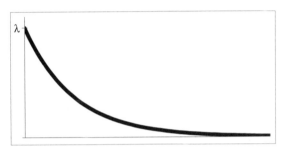

Figure 6.32

A random variable with an exponential distribution has mean = standard deviation = $\frac{1}{\lambda}$.

Example 6.5.4   Generating Values of a Random Variable

Once we know the density function $f(x)$ for a random variable X, we can generate values of it using the RAND function:

1. Find the CDF $y = F(x) = \int\limits_{-\infty}^{x} f(t)\, dt$.

2. Find the inverse of the CDF, $x = F^{-1}(y)$.

3. Use the RAND function to generate values of y and calculate values of X by $x = F^{-1}(y)$.

To illustrate this process, consider generating values of an exponentially distributed random variable X.

Step 1: Find the cumulative distribution function

$$F(x) = \int\limits_{-\infty}^{x} f(t)\, dt = \int\limits_{0}^{x} \lambda e^{-\lambda t} dt = 1 - e^{-\lambda x}.$$

Step 2: Set $F(x) = y$ and solve for x.

$$y = 1 - e^{-\lambda x} \quad \Rightarrow \quad x = -\frac{1}{\lambda}\ln(1-y) \quad \Rightarrow \quad F^{-1}(y) = -\frac{1}{\lambda}\ln(1-y)$$

Step 3: Let $y = \text{RAND}$, so the formula is

$$x = -\frac{1}{\lambda}\ln(1-\text{RAND}). \tag{6.3}$$

Formula (6.3) can be implemented in Excel very easily. Rename a blank worksheet "**Exponential**" and format it as in Figure 6.33. Copy row 3 down to row 1002 to generate a list of 1000 values of this random variable (we will use this list later).

	A	B
1		λ= 2
2	x	
3	=-1/$B$1*LN(1-RAND())	

Figure 6.33

Example 6.5.5   Generating Values of a Normal Random Variable

We can derive a formula for generating values of a normally distributed random variable in a similar fashion to Example 6.5.4. However, calculating F and F^{-1} is very complicated. Fortunately, Excel has F^{-1} built in for a normal distribution (and some other types of distributions as well). Rename a blank worksheet "**Normal**" and format it as in Figure 6.34. Copy row 4 down to row 1003 to generate 1000 values of this random variable.

	A	B
1		μ = 0
2		σ = 1
3	x	
4	=NORMINV(RAND(),$B$1,$B$2)	

Figure 6.34

Example 6.5.6   Generating Values of a Uniform Random Variable

A general uniformly distributed random variable can also be modeled easily in Excel. Rename a blank worksheet "**Uniform**" and format it as in Figure 6.35. Copy row 4 down to row 1003 to generate a list of 1000 values of this random variable.

	A	B
1		a = 5
2		b = 10
3	x	
4	=$B$1+($B$2-$B$1)*RAND()	

Figure 6.35

In the next section we verify (graphically at least) that these formulas do indeed generate numbers that fit the corresponding distributions.

Exercises

6.5.1 A random variable X has a density function given by

$$f(x) = \begin{cases} 1/(2\sqrt{x}) & \text{for } 0 < x < 1 \\ 0 & \text{elsewhere.} \end{cases}$$

1. Graph the density function $f(x)$ over the interval $[0,1]$.
2. Find the cumulative distribution function, $F(x)$, and graph it over the interval $[0,1]$.
3. Find the inverse cumulative distribution function, $F^{-1}(y)$, and use it to generate 100 values of X.

6.5.2 Repeat part (1), except use the density function

$$f(x) = \begin{cases} x^3/4 & \text{for } 0 < x < 2 \\ 0 & \text{elsewhere} \end{cases}$$

and graph over the interval $[0,2]$.

6.5.3 Repeat part (1), except use the density function

$$f(x) = \begin{cases} 2(1-x) & \text{for } 0 < x < 1 \\ 0 & \text{elsewhere.} \end{cases}$$

(**Hint:** Use the quadratic formula to solve for x in terms of y.)

6.6   Approximating Density Functions

In this section we illustrate how to graphically show that a particular density fits a given set of data. We start by showing that the 1000 values of the random variable we generated in the worksheet **Exponential** are indeed described by an exponential distribution.

When analyzing a large set of numerical data (the 1000 values we generated in this case), often the first step is to generate a *relative frequency histogram* such as that shown in Figure 6.36. To form a histogram like this, we first divide the overall range of data into several subintervals, or "bins." We then count the number of data values in each bin, and divide each count by the total number of data values to calculate the relative frequency of each bin. The histogram in Figure 6.36 shows, for instance, that almost all the data values are between 0 and 2, that about 33% of the data values are between 0 and 0.2, and that about 2% are between 1.4 and 1.6.

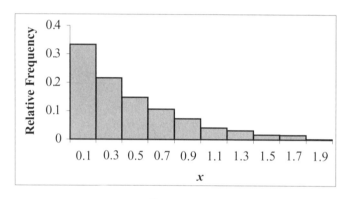

Figure 6.36

The first step in constructing a histogram is to determine an overall range of the data values. A good rule of thumb for any set of data is that the vast majority of all the data values will lie within three standard deviations of the mean. Since this data is supposedly exponential, the mean and standard deviation are both $1/\lambda$. So, since the data is positive, we consider the range

$$0 \text{ to } \frac{1}{\lambda} + 3\left(\frac{1}{\lambda}\right) = \frac{4}{\lambda}.$$

Next we need to decide on a number of bins. In this case we choose 10, but this number is somewhat arbitrary. (**Note:** There are several rules of thumb for determining the number of bins. One rule, called Sturgis's rule, is that the number of bins should be approximately $1 + 3.3 \log_{10} N$ where $N = $ the number of data values. There are also more complex ways of determining an "optimal" number of bins, but they are well beyond the scope of this book.) Then we need to calculate the width of each bin using

$$\text{Bin Width} = \frac{\text{Width of range}}{\text{Number of bins}}.$$

Add the formulas in Figure 6.37 to the worksheet **Exponential** created in Example 6.5.4 to calculate the bin width.

	D	E
1	# Bins =	10
2	Bin Width =	=(4/B1)/E1

Figure 6.37

Next, add the formulas in Figure 6.38 to count the number of data values in each bin and calculate the relative frequency of each. Copy row 6 down to row 14. The numbers in the column "Bins" are the midpoints of each of the bins. The formulas in column "Count" count the number of values of X in the interval

$$(\text{Bin} - 0.5b, \text{Bin} + 0.5b]$$

where $b = \text{Bin Width}$.

	C	D	E
4	**Bins**	**Count**	**Rel Frequency**
5	=0.5*E2	=COUNTIF($A$3:$A$1002,"<="&(C5+0.5*$E$2))	=D5/1000
6	=C5+$E$2	=COUNTIF($A$3:$A$1002,"<="&(C6+0.5*$E$2))-SUM($D$5:D5)	=D6/1000

Figure 6.38

Next we will create the relative frequency histogram. Follow these steps:

1. Click on the chart wizard (do not select any data yet).
2. Under **Chart type:** select **Column**. Under **Chart sub-type:** select the option in the upper-left corner. Click **Next**.
3. Under the **Series** tab, click **Add**. (If there is anything in the box next to **Data Range:** under the **Data Range** tab, delete it before clicking on the **Series** tab.)
4. For the **Values:**, select the range **E5:E14**.
5. For the **Category (x) axis labels:**, select the range **C5:C14**; click **Next**.
6. Name the **x-axis** "X" and the **y-axis** "Relative Frequency," and click **Finish.**
7. On the chart, set the y-axis **min** and **max** to **0** and **0.4**, respectively.
8. Right-click on a column in the chart, select **Format Data Series....** Under **Options**, set the **Gap width** to 0 and click **OK**.
9. Format the chart so it resembles Figure 6.36.

The leftmost column of the relative frequency histogram in Figure 6.36 shows the relative frequency of the values of X between 0 and 0.2. This is an approximation of $P(0 < X \leq 0.2)$. The column labeled 0.9 gives the relative frequency of the values of X between 0.8 and 1.0. This is an approximation of $P(0.8 < X \leq 1.0)$. As we can see,

$P(0 < X \leq 0.2) > P(0.8 < X \leq 1.0)$. This is what we expect if X is described by an exponential distribution.

Thus this relative frequency histogram gives us a way of graphically comparing the probabilities of X being in different intervals. The graph of the density function also does this. Therefore, if the density function $f(x) = \lambda e^{-\lambda x}$ really does describe this random variable, then the general shape of the relative frequency histogram should resemble the graph of $f(x)$. By examining Figure 6.36, we see that this is clearly true.

To better examine how closely the density function $f(x) = \lambda e^{-\lambda x}$ models this data, observe that in Figure 6.36,

$$
\begin{aligned}
0.33 &\approx P(0 < X \leq 0.2) \\
&= \int_0^{0.2} f(x)\, dx && \text{by definition} \\
&= 0.2 \cdot f(x_0) && \text{for some } x_0 \in [0, 0.2] \text{ by the mean value theorem.}
\end{aligned}
$$

Thus $0.2 \cdot f(x)$ should be approximately 0.33 over the interval $[0, 0.2]$. Graphically, this means that the graph of $y = 0.2 \cdot f(x)$ should intersect the top of the column in the histogram over the interval $[0, 0.2]$. This should be true for every other interval as well.

Therefore, if we graph $y = (\text{Bin Width}) \cdot f(x)$ on the same axis as the histogram, the curve should follow the tops of the columns. To do this, add the formulas in Figure 6.39 and copy row 5 down to row 14.

	F
4	**(Bin Width) * f(x)**
5	=$E$2*$B$1*EXP(-$B$1*C5)

Figure 6.39

To graph $y = (\text{Bin Width}) \cdot f(x)$ on top of the histogram, follow these steps:

1. Right-click on the histogram and select **Source Data. . . .**
2. Under **Series**, select **Add**.
3. For the **Values:**, select the range **F5:F14**.
4. For the **Category (x) axis labels:**, select the range **C5:C14** (this should be the default range).
5. Click **OK**.
6. On the chart, right-click on the Series 2 columns.
7. Select **Chart Type. . . .** Under **Chart sub-type:**, select **XY (Scatter)**, and select the **Chart sub-type:** on the middle right-hand side described as "Scatter with data points connected by smoothed lines without markers."
8. Your chart should resemble Figure 6.40.

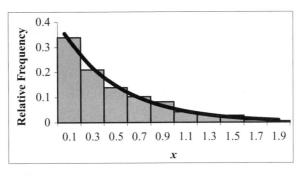

Figure 6.40

As we see, the graph of $y = (\text{Bin Width}) \cdot f(x)$ closely follows the tops of columns, thus graphically verifying that our formula generates values of an exponentially distributed random variable. Press **F9** several times to get different sets of data and try different values of λ. Note that in every case the graph closely follows the tops of the columns.

Example 6.6.1   Application to Real Data

Let's suppose we want to simulate the activity at a local grocery store checkout line. One random event we need to model is the arrival of customers. To do this, we observe 30 customers arriving at the checkout line and record the time between their arrivals (in minutes) as shown in Table 6.1.

1.40	2.79	0.91	1.87	0.87
1.60	1.76	3.46	1.51	3.90
0.03	0.36	0.21	2.36	3.24
0.21	0.10	0.96	0.92	0.47
5.75	0.90	0.66	1.56	2.74
0.22	1.33	0.04	0.33	1.84

Table 6.1

In our simulation, we need a density function to describe this random variable. Earlier we claimed that an exponential distribution is often used to describe the waiting time between events. We can use a modified version of the worksheet **Exponential** to test whether this is true in this case.

1. Create a copy of the worksheet **Exponential** and rename it "**Checkout Line.**" Type the data from Table 6.1 in the range **A3:A32**. Erase all entries below row 32.
2. Now we need to determine the value of the parameter λ. Note that for an exponential distribution, mean $= 1/\lambda$, so $\lambda = 1/\text{mean}$. To approximate the value of λ, add the formula shown in Figure 6.41. We see that $\lambda \approx 0.677$.

	A	B
1	λ =	=1/AVERAGE(A3:A32)

Figure 6.41

3. Next we need to modify the formula to calculate the relative frequencies as shown in Figure 6.42. Copy row 6 down to row 14.

	D	E
4	**Count**	**Rel Frequency**
5	=COUNTIF($A$3:$A$32,"<="&(C5+0.5*$E$2))	=D5/30
6	=COUNTIF($A$3:$A$32,"<="&(C6+0.5*$E$2))-SUM($D$5:D5)	=D6/30

Figure 6.42

The histogram is shown in Figure 6.43. Note that the graph follows the tops of the columns relatively well. This leads us to conclude that we can indeed model this random variable using an exponential density function with $\lambda = 0.677$.

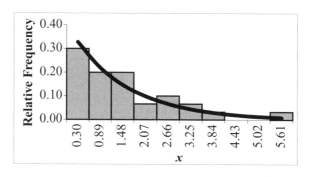

Figure 6.43

Exercises

6.6.1 Graphically verify that the formula used in the worksheet **Uniform** for generating values of a uniformly distributed random variable works for any appropriate values of a and b.

Suggestions

1. Use 10 bins.

2. Use the overall data range $[a, b]$ so that the Bin Width is $\frac{b-a}{10}$.

3. The first bin is at $a + 0.5$ (Bin Width).
4. Remember, the graph of the density function is a horizontal line.

6.6.2 Graphically verify that the formula used in the worksheet **Normal** for generating values of a normally distributed random variable works for any appropriate values of μ and σ.

Suggestions

1. Use 20 bins.
2. Use the overall data range $[\mu - 3\sigma, \mu + 3\sigma]$.

6.6.3 Suppose we want to simulate the packaging of "16-oz" packages of carrots. One random variable involved is the actual weight of the packages. We measure and record the weight (in ounces) of 30 packages as shown in Table 6.2.

15.99	16.18	16.16	16.38	16.10
16.28	16.34	16.20	16.21	16.27
16.26	16.10	15.94	16.52	16.16
16.38	16.11	16.43	16.21	16.36
16.08	16.15	16.31	16.06	16.06
16.01	16.26	16.19	16.30	16.08

Table 6.2

Graphically determine whether this random variable has a normal distribution. If so, estimate the mean μ and the standard deviation σ. (**Hint:** To calculate the standard deviation use the formula **STDEV.**)

6.7 A Theoretical Queuing Model

A *queue* is a waiting line. A prototypical example is people standing in line to buy movie tickets. Another example is an assembly line where bottles of soda wait to be filled with liquid. There are two important entities involved in a queue: *customers* and *servers*. Servers are whatever are used to process customers. In the movie ticket line, a customer is a person wanting to buy a ticket, and a server is a person selling tickets. There are three important parameters involved in describing a queue:

1. The number of servers.
2. Arrival rate—the average number of customers who arrive in the queue each time unit.
3. Service rate—the average number of customers each server can process each time unit.

Also important are the distributions that describe the rate at which customers arrive and the rate at which servers process customers (e.g., exponential, normal).

In this section we simulate a theoretical single-server queue with exponential arrival and server times (called an $M/M/1$ queue). This means that we model the *time between customer arrivals* (called the *inter-arrival time*) with an exponential distribution. We also model the *time to service customers* with an exponential distribution.

The queue system we look at has a mean arrival rate of $\lambda_1 = 2$ customers per minute and a mean service rate of $\lambda_2 = 3$ customers per minute. This means that the mean time between arrivals is $1/2$ min and the mean time to service a customer is $1/3$ min. We will use equation (6.3) to generate values of inter-arrival and service times.

Anyone waiting in line wants to know two things: (1) how long the line is and (2) how long he or she will have to wait to be serviced. Theorem 6.7.1 gives the answer (see Hillier and Lieberman 2001).

Theorem 6.7.1 *If the arrival rate is exponential and the service rate is given by any distribution, then the expected number of customers waiting in line (called the* expected queue length*),* L_q, *and the expected waiting time in the queue,* W_q, *are given by*

$$L_q = \frac{\lambda^2 \sigma^2 + \rho^2}{2(1-\rho)} \quad and \quad W_q = \frac{L_q}{\lambda}$$

where λ is the mean number of arrivals per time unit, μ is the mean number of customers serviced per time unit, $\rho = \frac{\lambda}{\mu}$, and σ is the standard deviation of the service time.

In this example, $\lambda = 2$ and since the service time is exponential, the standard deviation $\sigma = \text{mean} = 1/3$. So

$$L_q = \frac{4}{3} \quad and \quad W_q = \frac{2}{3}.$$

We will attempt to verify this value of W_q through our simulation.

Algorithm

1. For each customer:

    (a) Generate an inter-arrival time.
    (b) Calculate the arrival time.
    (c) Calculate the start time based on the finish time of the previous customer.
    (d) Generate the service time.

(e) Calculate the completion time.

(f) Calculate the amount of time spent waiting in line.

(g) Calculate the cumulative wait time for all customers up to this point.

(h) Calculate the average wait time for all customers up to this point.

2. Repeat for 5000 customers. This is one trial.

3. Repeat for 100 trials.

4. Find the overall average waiting time.

To implement this algorithm, follow these steps:

1. Rename a blank worksheet "**Queue.**" Format the worksheet as shown in Figure 6.44.

	A	B	C	D	E	F
1	**Mean Arrival Rate**	2	**Customer**	**Time Between**	**Arrival**	**Start**
2	**Mean Service Rate**	3	**Number**	**Arrivals**	**Time**	**Time**
3						
4			1	=-(1/$B$1)*LN(1-RAND())	=D4	=E4
5			=C4+1	=-(1/$B$1)*LN(1-RAND())	=E4+D5	=MAX(E5,H4)

Figure 6.44

2. Add the formulas in Figure 6.45. Copy the formulas in the range **C5:K5** down to row 5003 to simulate 5000 customers arriving.

	G	H	I	J	K
1	**Service**	**Completion**	**Wait**	**Cumulative**	**Average**
2	**Time**	**Time**	**Time**	**Wait Time**	**Wait Time**
3					
4	=-(1/$B$2)*LN(1-RAND())	=F4+G4	=F4-E4	=I4	=J4/C4
5	=-(1/$B$2)*LN(1-RAND())	=F5+G5	=F5-E5	=J4+I5	=J5/C5

Figure 6.45

3. Create a graph of Average Wait Time versus Customer Number as in Figure 6.46 (fix the y-axis **min** and **max** to **0** and **1.2**, respectively). The values on your graph may be different than in the figure due to the randomness of the simulation. Press **F9** several times to repeatedly simulate 5000customers arriving. Notice how much the average wait time varies as the number of customers increases. The average wait time for the very last customer is the overall average wait time. Notice that this number is not always very close to the theoretical average wait time of 0.666.

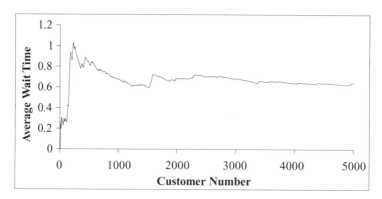

Figure 6.46

4. Next we need to create a table to store the results from 100 trials. Add the formulas in Figure 6.47. Copy row 4 down to row 102. Create a table in the range **M2:N102**. Choose any cell for the column input cell.

	M	N	O	P
1	**Trial Number**	**Avg. Wait Time**		**Overall Average**
2		=K5003		=AVERAGE(N3:N102)
3	1			
4	=M3+1			

Figure 6.47

Press **F9** several times and observe the overall average in cell **P2**. This should be close to the theoretical mean wait time. This example illustrates that it is often necessary to perform an experiment many, many times to get close to the theoretical average. This also illustrates that averages are not always representative of reality. The average wait time for the first 1000 customers or so (which would be a huge number of customers in a real situation of a checkout line) is not very close to the theoretical average and is often much larger. Because of this, one of the more important parts of queue design is to minimize standard deviations, not just averages, so that there are not as many extreme cases of persons waiting long periods of time.

Exercises

6.7.1 Change the mean arrival rate to 5 and the mean service rate to 6. Calculate the theoretical expected wait time, W_q, and verify this number with the simulation.

6.7.2 Suppose the mean arrival rate is 1 and the service time is uniformly distributed between 0 and 1 min. Calculate the theoretical expected wait time, W_q, and modify the formulas in your simulation to verify this number. (**Hint:** See Example 6.5.1 to calculate the mean and standard deviation of the service time. How many customers will the server be able to serve each minute? Find a formula to model the service time.)

6.7.3 The *queue length* is the number of customers standing in line waiting to be serviced when a customer arrives. This number does not include any customers being serviced at the time, nor does it include the arriving customer. Add a column to the worksheet **Queue** to calculate the queue length at the moment each customer arrives. Calculate the overall average queue length and compare it to the theoretical average $L_q = 4/3$ (use exponential distributions with a mean arrival rate of 2 and mean service rate of 3). (**Hint:** Consider the scenario shown in Figure 6.48. When customer 78 arrives, customer 76 is being serviced and customer 77 is waiting in line, so the queue length is 1. When customer 81 arrives, customer 77 is still being serviced, so customers 78, 79, and 80 are waiting in line. Use the COUNTIF function to count the number waiting in line.)

Customer Number	Arrival Time	Start Time	Completion Time	Queue Length
76	30.58	30.72	31.20	0
77	30.86	31.20	31.86	0
78	31.16	31.86	32.19	1
79	31.43	32.19	32.40	1
80	31.56	32.40	32.68	2
81	31.73	32.68	32.71	3

Figure 6.48

6.8 A Coffee Shop Queuing Model

Consider this scenario:

> RV Smith, the manager of RV's 66 gas station, has 100 regular customers who stop in for coffee between the hours of 6:00 and 7:30 AM. Currently he has only self-serve coffee with optional cream and sugar, but he is interested in opening an optional full-service gourmet coffee shop in the station.
>
> Talking with his customers, RV learns that about 80% of them would be interested in using the full-service shop. However, they say they won't wait more than about 2 min for their coffee. This is a concern to RV since he will be the only one working in the shop and he estimates he can serve about one person per minute.

RV wants to estimate the number of customers he will be able to serve during this time span so he can determine if it is worthwhile to open the shop. RV has observed that the rate at which his coffee customers arrive steadily increases from 6:00 to 6:15, then remains steady from 6:15 to 6:45, and then steadily decreases from 6:45 to 7:30.

In our simulation we will consider only the 100 regular coffee customers and the time from 6:00 to 7:30 AM. We will assume that the service time (i.e., the time to service the full-service coffee) is exponentially distributed. We will also assume that the probability of any one person wanting full-service coffee is 0.80. Note that this scenario is a queue, but it is much more complicated than the theoretical scenario considered earlier. The inter-arrival times cannot be modeled with a simple density function, so trying to solve this problem theoretically would be extremely difficult. One great benefit of simulation is that it allows us to model complicated scenarios such as this one relatively simply.

The basic algorithm behind the simulation will be similar to the theoretical queuing model. We will generate an arrival time for each of the 100 customers. However, not everyone will actually get in line at the coffee shop. Also, people will get out of line if they have to wait more than 2 min.

The biggest difference is in how we generate arrival times. Instead of generating inter-arrival times using an exponential distribution, we will generate arrival times using the description of the rate of customer arrivals.

If the rate at which customers arrive steadily increases from 6:00 to 6:15, then there will be relatively more customers arriving near 6:15 than near 6:00. If the rate is steady between 6:15 and 6:45, then there will be relatively the same number of customers who arrive near 6:15 as near 6:45. If the rate is steadily decreasing between 6:45 and 7:30, then there will be relatively more customers who arrive near 6:45 than near 7:30. Thus a relative frequency histogram of the arrival times will look similar to Figure 6.49.

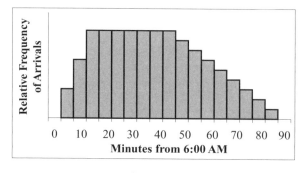

Figure 6.49

Let the random variable X represent arrival time (in minutes from 6:00 AM). This random variable has a density function whose graph resembles that in Figure 6.50.

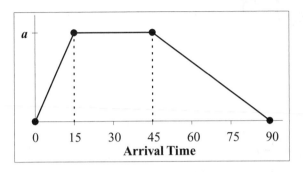

Figure 6.50

We will use this density function to calculate values of X with the process described in Example 6.5.4. We first need to find the value of a. We know that for any density function, the area under the curve must be exactly 1. This means, in Figure 6.50,

Area of left triangle + area of center rectangle + area of right triangle = 1

$$\Rightarrow \quad \tfrac{1}{2}(15)\, a + 30a + \tfrac{1}{2}(45)\, a = 1 \quad \Rightarrow \quad a = \tfrac{1}{60}.$$

Thus the density function is given by

$$f(x) = \begin{cases} 0 & x < 0 \\ (1/900)\, x & 0 \le x < 15 \\ (1/60) & 15 \le x < 45 \\ -(1/2700)\, x + (1/30) & 45 \le x < 90 \\ 0 & 90 \le x. \end{cases}$$

Next we need to calculate the CDF, $F(x)$. For $0 \le x < 15$,

$$F(x) = \int_0^x \frac{1}{900} t\, dt = \frac{1}{1800} t^2 \Big|_0^x = \frac{x^2}{1800}$$

for $15 \le x < 45$,

$$F(x) = \frac{15^2}{1800} + \int_{15}^x \frac{1}{60}\, dt = \frac{1}{8} + \frac{1}{60}(x - 15) = \frac{x}{60} - \frac{1}{8}$$

and for $45 \leq x < 90$,

$$F(x) = \left(\frac{45}{60} - \frac{1}{8}\right) + \int_{45}^{x} -\frac{1}{2700}t + \frac{1}{30}dt = \frac{5}{8} - \frac{t^2}{5400} + \frac{t}{30}\bigg|_{45}^{x} = -\frac{x^2}{5400} + \frac{x}{30} - \frac{1}{2}.$$

Thus the CDF is:

$$F(x) = \begin{cases} 0 & x < 0 \\ (x^2/1800) & 0 \leq x < 15 \\ (x/60) - (1/8) & 15 \leq x < 45 \\ -(x^2/5400) + (x/30) - (1/2) & 45 \leq x < 90 \\ 1 & 90 \leq x. \end{cases}$$

The graph of $F(x)$ is shown in Figure 6.51.

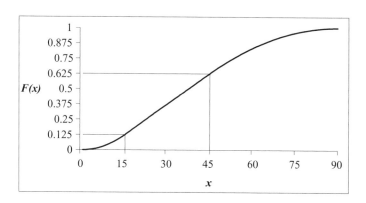

Figure 6.51

Next we need to find the inverse CDF, $F^{-1}(x)$. For $0 \leq y < 0.125$,

$$y = \frac{x^2}{1800} \quad \Rightarrow \quad x = \sqrt{1800y}$$

for $0.125 \leq y < 0.625$,

$$y = \frac{x}{60} - \frac{1}{8} \quad \Rightarrow \quad x = 60\left(y + \frac{1}{8}\right)$$

and for $0.625 \leq y < 1$,

$$y = -\frac{x^2}{5400} + \frac{x}{30} - \frac{1}{2} \quad \Rightarrow \quad x = 90 - \sqrt{5400\,(1 - y)}.$$

Thus the inverse CDF is

$$F^{-1}\,(y) = \begin{cases} \sqrt{1800y} & 0 \leq y < 0.125 \\ 60\,(y + 0.125) & 0.125 \leq y < 0.625 \\ 90 - \sqrt{5400\,(1 - y)} & 0.625 \leq y < 1. \end{cases} \qquad (6.4)$$

Algorithm

1. Generate the arrival times using (6.4) for 100 customers.
2. Sort the arrival times in ascending order.
3. For each customer in the order in which he or she arrived:

    (a) Determine if he or she wants full service.
    (b) Determine if he or she would have to wait too long based on the completion time of the previous customer.
    (c) Determine a start time.
    (d) Calculate a service time using an exponential distribution.
    (e) Determine if he or she actually received full service based on (a) and (b).
    (f) Determine his or her completion time. If he or she did not receive full service, his or her completion time is the completion time of the previous customer.

4. Calculate the actual number of customers served.

To implement this algorithm, follow these steps:

1. Rename a blank worksheet "**Coffee Shop**," format it as in Figure 6.52, and copy row 8 down to row 107. Here, x represents the arrival time of the customer in minutes from 6:00 AM.

	A	B	C	D
5			x	Arrival
6	#	rand		Time
7	0			
8	1	=RAND()	=IF(B8<0.125,SQRT(1800*B8),IF(B8<0.625,60*(B8+0.125),90-SQRT 5400*(1-B8))))	

Figure 6.52

2. In column **D** we sort the arrival times in column **C** in ascending order. Specifically, we copy the times in column **C**, paste their values in column **D**, and then sort the

values in column **D**. When we do this, we record our steps in the form of a *macro* so we can easily repeat them. Follow these steps *very carefully*. Excel is recording everything you do:

(a) Select **Tools → Macros → Record New Macro....** Name the Macro "**Arr Times.**" Click **OK**.

(b) Press the **F9** key.

(c) Highlight the range **C8:C107** and copy it.

(d) Select the cell **D8** and select **Edit → Paste Special....** Select **Values** and click **OK**.

(e) Select **Data → Sort...**.

(f) Select **Continue with the current selection** and click **Sort....** Click **OK**.

(g) Press the **F9** key again.

(h) Click the small square in the Macro control window to stop recording.

(i) Select **View → Toolbars → Forms**.

(j) In the Forms window, select the second option on the right-hand side, **Button**.

(k) Draw a button anywhere on the worksheet.

(l) Select the Macro **ArrTimes** and click **OK**.

(m) Rename the button "**Run.**" New arrival times will be generated and sorted each time you click this button. This simulates the arrival of the 100 customers.

(n) Close the Forms window.

3. Add the formulas in Figure 6.53 to store the values of the parameters and display the results.

	B	C
1	# Customers Serviced =	=SUM(I8:I107)
2	Mean Service Rate =	1
3	Max Wait Time =	2
4	% Want Full Service =	0.8

Figure 6.53

4. Add the formulas in Figure 6.54 to determine whether each customer wants full service, if he or she would have to wait too long, and the start time. The formulas in column **F** will return a 1 if the customer waited too long and a 0 otherwise. Copy row 8 down to row 107.

F	G
Wait Too Long?	**Start Time**
=IF(AND(J7-D8>$C$3,E8=1),1,0)	=MAX(D8,J7)

Figure 6.54

to calculate the service and completion time for each
er the customer actually received full service. The
n a 1 if the customer received full service and a 0
row 107.

I	J
Received Full Service?	**Completion Time**
	0
=IF(AND(E8=1,F8=0),1,0)	=IF(I8=1,G8+H8,J7)

Figure 6.55

graph of Waited Too Long? versus Arrival Time as in Figure 6.56.

Click the **Run** button several times and note the number of customers serviced. The
number is typically between 55 and 60. This means that RV can expect to serve
about 55 to 60 customers at the full-service shop each morning. Also note the arrival
times of customers who waited too long. Most are between the time of 15 and 45,
as expected, as this is the busiest time of the morning. However, there are some
customers who arrive during the slow periods who waited too long.

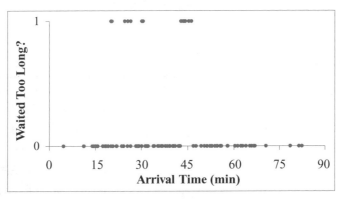

Figure 6.56

Exercises

6.8.1 Investigate the sensitivity of the system to the value of the parameter "Mean Service Rate." That is, change its value and determine how much the number of customers serviced changes.

1. If RV were to break his arm and not be able to serve coffee as quickly, how would this affect business?

2. If RV were to hire an additional person to work in the shop with him and double the rate at which customers were served, how would this affect business?

3. What if he were to hire two additional people to work with him and triple the service rate? Would this be advantageous over hiring just one additional person?

6.8.2 Investigate the sensitivity of the system to the value of the parameter "% Want Full Service" (keep the Mean Service Rate at 1).

1. If a higher percentage of persons started wanting full-service coffee, how many more persons would RV be able to serve?

2. If a lower percentage wanted full service, would it affect business dramatically?

6.8.3 Suppose the arrival times are uniformly distributed between 6:00 and 7:30. How many customers would RV be able to serve? How does this affect the times at which customers waited too long?

6.8.4 Suppose that customers start arriving at a constant rate between 6:00 and 7:00 but then the rate doubles between 7:00 and 7:30. How many customers would RV be able to serve? How does this affect the times at which customers waited too long?

6.8.5 Suppose that the rate at which customers arrive steadily decreases between 6:00 and 7:00 so the rate is near 0 at 7:00, but then the rate steadily increases from 7:00 to 7:30 so that the rate near 7:30 is the same as the rate near 6:00. How many customers would RV be able to serve? How does this affect the times at which customers waited too long?

6.9   A Scheduling Model

The Handyman Remodeling Company is considering placing a bid on a contract to remodel a living and dining room. The contract includes a large penalty if the job is not completed by the deadline of 3 weeks (21 days) from the start. The job would consist of eight separate tasks with various precedence constraints as illustrated in Figure 6.57.

For each task, the project manager has estimated a *most likely* duration, m, an *optimistic* duration, a, and a *pessimistic* duration, b, in terms of the number of days. These numbers are given in Table 6.3.

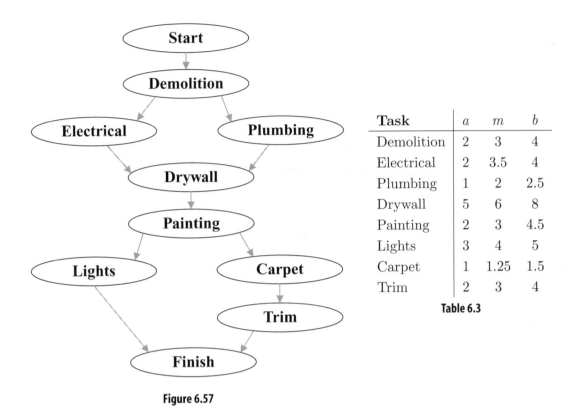

Task	a	m	b
Demolition	2	3	4
Electrical	2	3.5	4
Plumbing	1	2	2.5
Drywall	5	6	8
Painting	2	3	4.5
Lights	3	4	5
Carpet	1	1.25	1.5
Trim	2	3	4

Table 6.3

Figure 6.57

Company managers want to know the expected project duration and the probability of meeting the deadline so they can decide whether to place the bid. We will create a simulation to help answer this question.

The duration of each task is a random variable with a distribution. The problem is that we do not know what the distribution is, and there is no reasonable way to determine what the density function is by collecting data. Therefore, the best we can do is use a reasonable distribution to model these random variables. The distribution we will use is called a *triangular distribution*. Its density function is graphed in Figure 6.58.

If we model the duration of a task with a triangular distribution, we see in Figure 6.58 that the probability the duration is near m is much higher than the probability it is near the optimistic or pessimistic estimate. This observation indicates that the triangular distribution is indeed a reasonable model.

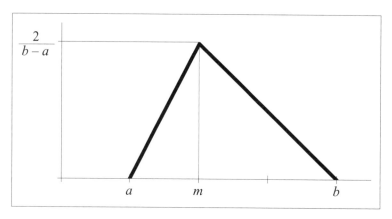

Figure 6.58

It can be shown that the CDF for a triangular distribution is given by

$$F(x) = \begin{cases} \frac{(x-a)^2}{(m-a)(b-a)} & a \le x \le m \\ 1 - \frac{(b-x)^2}{(b-m)(b-a)} & m < x \le b \end{cases}$$

and the inverse CDF, $F^{-1}(x)$, is given by

$$F^{-1}(y) = \begin{cases} a + \sqrt{y(m-a)(b-a)} & 0 \le y \le \frac{m-a}{b-a} \\ b - \sqrt{(1-y)(b-m)(b-a)} & \frac{m-a}{b-a} < y \le 1. \end{cases}$$

Algorithm

1. For each task:

   (a) Determine the start time based on the finish times of the preceding activities.
   (b) Generate the task duration.
   (c) Calculate the finish time.

2. Determine the overall project duration.

3. Determine whether the deadline was met.

4. Repeat for 500 trials.

5. Calculate the average project duration and the percentage of the trials in which the deadline was met.

To implement this algorithm, follow these steps:

1. Rename a blank worksheet "**Remodel**" and format it as in Figure 6.59.

	A	B	C	D	E	F	G	H
1	**Task**	**a**	**m**	**b**	**Start Time**	**rand**	**Duration**	**Finish Time**
2	Demo	2	3	4	0	=RAND()		=E2+I16
3	Electrical	2	3.5	4	=H2	=RAND()		=E3+I17
4	Plumbing	1	2	2.5	=H2	=RAND()		=E4+I18
5	Drywall	5	6	8	=MAX(H3,H4)	=RAND()		=E5+I19
6	Painting	2	3	4.5	=H5	=RAND()		=E6+I20
7	Lights	3	4	5	=H6	=RAND()		=E7+I21
8	Carpet	1	1.25	1.5	=H6	=RAND()		=E8+I22
9	Trim	2	3	4	=H8	=RAND()		=E9+I23
10								
11							**Project Duration =**	=MAX(H7,H9)
12							**Deadline Met?**	=IF(H11<=21,1,0)

Figure 6.59

2. Add the formula for the inverse CDF as shown in Figure 6.60, and copy cell **G2** to the range **G3:G9**.

	G
2	=IF(F2<=(C2-B2)/(D2-B2),B2+SQRT (F2*(C2-B2)*(D2-B2)),D2-SQRT((1-F2)*(D2-C2)*(D2-B2))

Figure 6.60

3. Start a table to store the results of 500 trials and find the average duration and percentage of successes as shown in Figure 6.61. Copy cell **J4** down to row 502. Create a table in the range **J2:L502** to store the results of 500 trials. Choose any blank cell for the **Column input cell**.

	I	J	K	L
1	**Average Duration**	**Trial**	**Project Duration**	**Success**
2	=AVERAGE(K3:K502)		=H11	=H12
3	**% Success**	1		
4	=AVERAGE(L3:L502)*100	=J3+1		

Figure 6.61

From the results we see that there is approximately an 81% chance of finishing on time and the average duration is just over 20 days.

Exercises

6.9.1 The quantity "project duration" is a random variable.

1. Approximate the mean and standard deviation of this random variable.

2. Create a relative frequency histogram of the 500 values of this variable generated in the simulation, and determine which type of density function (i.e., uniform, exponential, or normal) best models this random variable.
3. Fit an appropriate density curve to the relative frequency histogram and comment on how well it fits.

6.9.2 Instead of modeling the duration of each task as a continuous random variable over the interval $[a, b]$, model it as a *discrete* random variable that will take a value of a, m, or b, each with a certain probability. For instance, suppose it equals a with probability 0.25, m with probability 0.5, and b with probability 0.25. Modify the worksheet from Exercise 6.9.1 to model the durations in this way. How does this change the mean, standard deviation, and distribution of the variable project duration? Try different values of the probabilities.

6.9.3 Graphically verify that our Excel formula for generating values of a random variable X described by a triangular distribution really does work with $a = 3$, $m = 5.25$, and $b = 9$. (**Suggestions:** Use 12 bins in your histogram and a bin width of 0.5. The first bin is centered at 3.25.)

6.9.4 Martin has 4 days to study for his Mathematical Modeling final exam. At the end of day 0 he has an entire 100 pages of notes to read. He figures that if he spends h hours studying on any given day he will absorb $11.12\sqrt{h}$ pages of notes. He also figures that overnight he'll forget 10% of what he knew at the end of that day. So if we let h_i = the number of hours spent studying on day i and x_i = the number of pages absorbed by the end of day i, we have $x_0 = 0$ and

$$x_i = 0.9x_{i-1} + 11.12\sqrt{h_i}.$$

Martin would like to go into the exam having absorbed everything (i.e., $x_4 = 100$), and has planned to study a certain number of hours each day. However, he realizes that he may get lazy and might not be able to study as many hours as he planned. Conversely, he may get very motivated and study more hours than anticipated. Therefore, for each day he has estimated a *most likely* study duration, an *optimistic* duration, and a *pessimistic* duration as shown in Table 6.4. Design a simulation to estimate the probability that he will absorb everything and the expected number of pages he will absorb. Should Martin revise his time estimates? If so, how?

Day	Pessimistic	Most Likely	Optimistic
1	2	4	6
2	5	6	7.5
3	6	7	9
4	7.5	9	11

Table 6.4

6.10   An Inventory Model

Consider the following scenario:

The produce department at a neighborhood grocery store gets its bananas from a local supplier. To better schedule its deliveries, the supplier has asked the produce manager to place his order on a regular basis (e.g., every 5 days). The manager is trying to decide how often to place his order.

The manager has room to store 50 boxes of bananas. Each time he places an order, he orders enough to completely replenish his stock. He will place his order at the end of the day and it will be delivered that evening. There is a $25 delivery fee for each delivery, regardless of the size of the order.

The manager looks over his records from the previous month (30 days) for daily demand and notes that he sold between 1 and 10 boxes of bananas each day. The daily demand data is summarized in Table 6.5.

Demand	1	2	3	4	5	6	7	8	9	10
Number of Days	3	3	2	2	3	5	2	4	3	3

Table 6.5

Ideally, the manager wants to place his order before he runs out of bananas. For instance, if at the end of day 8 he has 2 boxes left and on day 9 he has demand for 5 boxes, he would have wished he had ordered 48 boxes on day 8. However, he doesn't want to place an order too often because of the delivery fee, so he is willing to occasionally run out of bananas.

The above scenario is somewhat vague. So, we first need to state a precise question to answer:

How often should the manager place his order so that the produce department can meet demand at least 95% of the time?

Our simulation will consist of replicating the sale and delivery of bananas over a period of 1 year. We will vary the number of days between deliveries and keep track of the number of days demand was met.

The only random variable in this simulation is the daily demand. This is a discrete random variable and we will use the data in Table 6.5 to model the CDF. To do this, rename a blank worksheet "**Bananas**" and format it as in Figure 6.62. Enter the rest of the data from Table 6.5 and copy the range **O4:P4** down to row 12.

The graph of the cumulative distribution function, $F(x)$, is shown in Figure 6.63. (**Note:** It is not necessary to create this graph.) To generate values of the demand based on this CDF we first pick a uniformly distributed random number y using the RAND function. From that y-value on the graph, we will move horizontally until we hit a column. The

	M	N	O	P
1		Num of	Relative	Cumulative
2	Demand	Days	Frequency	Frequency
3	1	3	=N3/30	=O3
4	2	3	=N4/30	=O4+P3

Figure 6.62

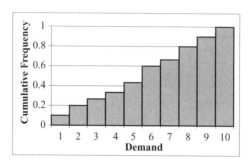

Figure 6.63

corresponding x-value will be the value of $F^{-1}(y)$. For instance, $F^{-1}(0.065) = 1$ and $F^{-1}(0.75) = 8$.

To calculate $F^{-1}(y)$, add the formulas in Figure 6.64 and copy row 4 down to row 12. The simulation will refer to the "lookup chart" to calculate the daily demand.

	R	S
1	Lookup	Chart
2	Frequency	Demand
3	0	1
4	=P3	2

Figure 6.64

Algorithm

1. Choose a value for days between deliveries.
2. For each day:

    (a) Generate a demand.
    (b) Determine whether the demand can be met.
    (c) Determine whether a delivery will be made.
    (d) Calculate inventory at the end of the day.

3. Calculate the percentage of days the demand was met.

4. Repeat for 100 trials.

5. Try different values of days between deliveries.

To implement this algorithm, follow these steps:

1. In the worksheet **Bananas**, add the formulas in Figure 6.65. Copy row 6 down to row 370. The **VLOOKUP** function in column **C** will look at the leftmost column in the lookup chart, find the largest value less than or equal to the value of the random number, and return the corresponding value in the second column of the chart. This value will be the demand for that day.

	A	B	C	D
1			**Days Between Deliveries =** 5	
2				
3				
4	**Day**	**rand**	**Demand**	**Delivery?**
5	0			
6	=A5+1	=RAND()	= VLOOKUP(B6,$R$3:$S$12,2)	=IF(MOD(A6,$D$1)=0,1,0)

	E	F
3	**Demand**	**Inventory**
4	**Met?**	**at End of Day**
5		50
6	=IF(F5>=C6,1,0)	=IF(D6=1,50,F5-C6)

Figure 6.65

2. Add the formulas in Figure 6.66 to calculate the percentage of days that demand was met, store the results from 100 trials, and calculate the overall results. Create a table in the range **I4:J104** to store the results from 100 trials. Choose any blank cell for the column input cell. Press **F9** several times to repeat the simulation. Note that with 5 days between deliveries, demand is met 100% of the time.

	I	J
1	**% Days Demand Met =**	=AVERAGE(E6:E370)*100
2	**Overall Average =**	=AVERAGE(J5:J104)
3	**Trial**	**%**
4		=J1
5	1	
6	=I5+1	

Figure 6.66

3. Change the number of days between deliveries and rerun the simulation until you find the largest value that gives at least a 95% overall average. Note that for 8 days

between deliveries demand is met about 95.5% of the time. For 9 days between deliveries demand is met only about 90% of the time. Thus 8 days between deliveries is the answer to the question. In the exercises we will consider a refinement to the model.

Exercises

6.10.1 Use the data in Table 6.5 to calculate the expected daily demand. Remember, for a discrete random variable X, $E(X) = \sum_{x=-\infty}^{\infty} x f(x)$ where $f(x)$ = the probability of observing the value of x (the relative frequency in this case).

6.10.2 Add a formula to the simulation to calculate the average daily demand for a trial. Store this average in the table and calculate an overall average for the 100 trials. Does this overall average agree with the expected value calculated in Exercise 6.10.1? Does our method for generating values of the random variable daily demand appear to work properly?

6.10.3 Suppose that the produce department makes $3 profit for each box of bananas sold (if we don't take into account delivery costs).

1. Add a column of formulas to calculate daily profit. (**Note:** The daily profit is the number of boxes *sold* times $3, not the number *demanded* times $3.)

2. Add a formula to calculate the total delivery cost for the year (each delivery costs $25).

3. Add a formula to calculate the total yearly profit from the sale of bananas.

4. Modify your table to store the profit and add a formula to calculate the average profit from all 100 trials.

5. Find the number of days between deliveries that gives the maximum profit. How does this value compare to the solution to the original problem of 8 days between deliveries?

For Further Reading

- For a classic reference on many of the concepts related to simulation, see F. Hillier and G. Lieberman, 2001, *Introduction to Operations Research*, 7th ed., McGraw-Hill, Boston, MA, pp. 1084–1155.

- For a classic reference on everything related to simulation, see A. M. Law and W. D. Kelton, 1991, *Simulation Modeling and Analysis*, 2nd ed., McGraw-Hill, New York, NY.

- For more examples of the concepts in this chapter, see D. Maki and M. Thompson, 2006, *Mathematical Modeling and Computer Simulation*, Thomson Brooks/Cole, Belmont, CA.

References

Giordano, F. R., M. D. Weir, and W. P. Fox. 2003. *A first course in mathematical modeling*, 3rd ed. Pacific Groves, CA: Thomson Brooks/Code, 182.

Hiller, F., and G. Lieberman, 2001. *Introduction to operations research*, 7th ed. Boston, MA: McGraw-Hill, 871, 1087–1088. Reproduced by permission of McGraw-Hill Companies.

Microsoft Help and Support Webpage. *http://support.microsoft.com/kb/q86523/*. Accessed June 2008.

Microsoft Help and Support Webpage. *http:/support.microsoft.com/default.aspx?scid=kb; en-us;828795*. Accessed June 2008.

CHAPTER 7

Optimization

Chapter Objectives

- Discuss the basic concepts of optimization problems
- Introduce linear programming
- Model transportation and assignment problems
- Discuss the basics of the simplex method
- Introduce nonlinear programming

7.1   Introduction

In Calculus I, we solved problems such as

$$\text{Maximize} f(x) = -2x^2 + 3x + 2$$

by taking the derivative of the function f and setting it equal to 0. This is a very simple *optimization problem* (OP). Practical situations often involve finding solutions to more complex optimization problems such as when a business is trying to decide how many units of a product to produce so as to maximize profit.

Every OP has two components: an *objective function* and *decision variable(s)*. The objective function is the function being maximized. The decision variable(s) are the variable(s) involved. The basic goal of an OP is to find values of the decision variable(s) that maximize the objective function.

Optimization problems are classified into two general categories: *constrained* and *unconstrained*. A constrained OP is one in which there are constraints on the values of the decision variable(s). An unconstrained OP has no such constraints. These constraints can be of many different forms, including:

1. Nonnegativity (the decision variables must be nonnegative)
2. Integrality (the decision variables must be integers)
3. Binary (the decision variables must be 0 or 1)
4. Equality (e.g., $x + y = 5$)
5. Inequality (e.g., $x + y \geq 6$)

Optimization problems are also classified into two other categories: *linear* or *nonlinear*. A linear OP is one in which the objective function and constraints are equations or inequalities of the form

$$2x_1 - 6x_2 + 8x_3 = 2$$

where x_1, x_2, and x_3 are decision variables (i.e., the objective function and constraints are equations or inequalities of the type studied in linear algebra). A nonlinear OP has an objective function or at least one constraint that is not of this type.

Of course, OP could involve minimizing a function. If the objective is to minimize $g(x)$ subject to some set of constraints, we could transform this into an equivalent maximization problem:

$$\text{Maximize } -g(x)$$

subject to the same constraints. Therefore, we mostly consider maximization problems in this chapter.

7.2 Linear Programming

The type of optimization problem we spend most time on is called a *linear program* (LP), which is simply a linear constrained OP. A typical linear program has the form:

$$\text{Maximize } Z = 25x_1 + 30x_2$$
$$\text{Subject to} \quad 20x_1 + 30x_2 \leq 600$$
$$5x_1 + 4x_2 \leq 4 \quad\quad (7.1)$$
$$x_1 \geq 4, x_2 \geq 2.$$

The objective function in this case is $25x_1 + 30x_2$, and its value is denoted by Z. A *solution* to an LP (sometimes called a *schedule*) is any value of the decision variables that satisfies the constraints. A solution to the LP (7.1) is $x_1 = 5$, $x_2 = 6$. An *optimal solution* is one that gives the largest value of the objective function over all possible solutions. Finding solutions is relatively easy. Showing that a particular solution is an optimal solution is not as easy.

The algorithm used to find optimal solutions to linear programming problems is called the *simplex method*. Excel uses this algorithm as part of its **Solver** tool.

Example 7.2.1 Making Fruit Baskets

The manager of a produce department at a neighborhood supermarket is making fruit baskets for the busy holiday season. He sells two sizes of baskets: small and large. He has only 200 apples and 100 oranges remaining and is trying to decide how many of each size of basket he should make. Each small basket returns a profit of $3 and requires 3 apples and 1 orange while each large basket returns a profit of $4 and requires 2 apples and 2 oranges. Assuming he will sell all that he makes, how many baskets of each size should he make to maximize his profit?

The first step in modeling a problem such as this is to organize all the information given into a *mixture chart* as shown in Table 7.1. The two sizes of baskets are generically called the *products* and the apples and oranges are called the *resources*.

	Products		
Resources	Small	Large	Amount Available
Apples	3	2	200
Oranges	1	2	100
Profit	3	4	

Table 7.1

The next step is to define variables and write a set of inequalities to model the situation. If $s =$ the number of small baskets to produce and $l =$ the number of large baskets to produce, our model is

$$\text{Maximize } P = 3s + 4l$$
$$\text{Subject to} \quad 3s + 2l \leq 200$$
$$1s + 2l \leq 100$$
$$s, l \geq 0.$$

The objective function $3s + 4l$ gives the total profit. The first constraint says we cannot use more than 200 apples while the second says we cannot use more than 100 oranges. The last two constraints are nonnegativity constraints that say we cannot produce a negative number of baskets.

To solve this LP, rename a blank worksheet "**Fruit Baskets**" and format it as in Figure 7.1. The numbers in the range **B2:C2** are the values of s and l (these are not yet the optimal values).

	A	B	C	D	E
1		Small	Large		
2	**Number**	1	1	**Amt Used**	**Amt Available**
3	**Apples**	3	2	=SUMPRODUCT($B$2:$C$2,B3:C3)	200
4	**Oranges**	1	2	=SUMPRODUCT($B$2:$C$2,B4:C4)	100
5				**Total Profit**	
6	**Profit**	3	4	=SUMPRODUCT($B$2:$C$2,B6:C6)	

Figure 7.1

Select **Tools** → **Solver** ... (if **Solver** ... is not available, select **Tools** → **Add–Ins** ... , select **Solver Add-in**, and click **OK**). Format the Solver window as in Figure 7.2.

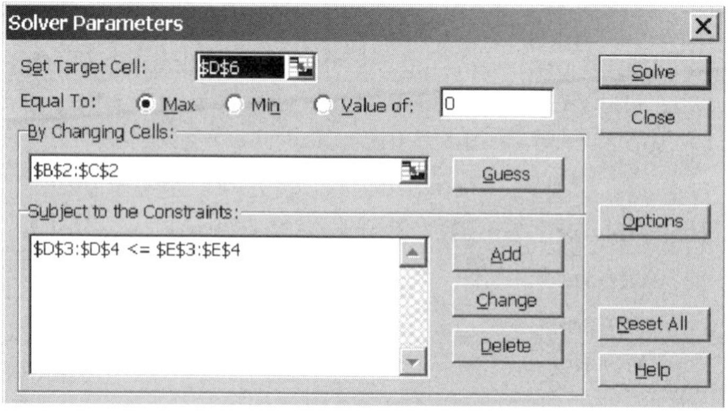

Figure 7.2

In the Solver window, click **Options**, and select **Assume Linear Model** and **Assume Non-Negative**. Click **OK**, then **Solve**, and then **OK**. Your worksheet should now look like Figure 7.3. From this we see that our optimal solution is to produce 50 small baskets and 25 large baskets (called the optimal *production schedule*), which will yield a profit of $250. Also note that we use all 200 apples and 100 oranges.

	A	B	C	D	E
1		Small	Large		
2	**Number**	50	25	**Amt Used**	**Amt Available**
3	**Apples**	3	2	200	200
4	**Oranges**	1	2	100	100
5				**Total Profit**	
6	**Profit**	3	4	250	

Figure 7.3

Example 7.2.2   A Diet

John has decided to go on a diet to lose weight and has limited himself to two types of food: protein shakes and pasta (plus vitamin supplements). He is concerned with getting enough protein and carbohydrates, but not too much fat. Table 7.2 lists the amounts of these nutrients provided by each type of food along with his daily requirements and cost information. How many servings of each food should John eat each day to meet the requirements and minimize the cost?

	Grams per Serving		
	Shake	Pasta	Daily Requirement
Carbs	1	35	≥ 80
Protein	7	6	≥ 75
Fat	5	1	≤ 50
Cost per Serving	$0.75	$1.25	

Table 7.2

If we let s = the number of protein shake servings and p = the number of pasta servings per day, our mathematical model is

$$\text{Minimize } C = 0.75s + 1.25p$$
$$\text{Subject to} \quad 1s + 35p \geq 80$$
$$7s + 6p \geq 75$$
$$5s + 1p \leq 50$$
$$s, p \geq 0.$$

Note that in this case, the objective function gives the cost, not the profit. Solving this problem in a similar fashion to the previous example gives the results shown in Figure 7.4. They indicate that John should eat about nine shakes and two servings of pasta each day with a daily cost of about $9.25.

	A	B	C	D	E
1		Shake	Pasta		
2	Number	8.975	2.029	Amt Consumed	Amt Needed
3	Carbs	1	35	80	80
4	Protein	7	6	75	75
5	Fat	5	1	46.90376569	50
6				Total Cost	
7	Cost	0.75	1.25	9.267782427	

Figure 7.4

Exercises

Directions: Formulate each problem below as an LP and solve it with Solver.

7.2.1 A toy company manufactures plastic cars and trucks. Each car yields a profit of $1.25 and requires 5 units of plastic and 15 min of labor to produce. Each truck yields a profit of $0.95 and requires 2 units of plastic and 18 min of labor to produce. If the company has 60 units of plastic and 360 min of labor available, how many of each vehicle should it produce to maximize its profit?

7.2.2 The Sweetie Pie Baking Company operates three different plants. Two are used for mixing ingredients and baking, and the third is used strictly for packaging. Management is considering adding two new products to their line-up: a chocolate-chunk cookie and a raisin bread. The cookie will be baked in plant 1 while the bread will be baked in plant 2. They will both be packaged in plant 3. One batch of each product requires a certain amount of time in the different plants and each plant has a certain amount of production time available each week. The data is summarized in Table 7.3. How many batches of each product should be produced each week to maximize total profit?

	Production Time Needed		Production Time
	Cookie	Bread	Available per Week
Plant 1	3	0	6
Plant 2	0	2	10
Plant 3	1	2	10
Profit per Batch	300	250	

Table 7.3

7.2.3 The Nutty Goodness Company sells mixtures of peanuts, walnuts, and cashews. A new customer wants 100 lb of a mixture that is 45% peanuts, 30% walnuts, and 25% cashews. The company has run out of peanuts, so it is going to make the mixture by combining five different mixtures containing different percentages of peanuts, walnuts, and cashews as shown in Table 7.4. Determine the amounts of each of the five different mixtures that should be blended to form 100 lb of the new mixture at a minimum cost.

	Mixture				
	1	2	3	4	5
Percentage of peanuts	45	20	55	50	55
Percentage of walnuts	23	20	42	20	35
Percentage of cashews	32	60	3	30	10
Cost ($/lb)	4.80	5.20	4.90	4.60	4.30

Table 7.4

7.3   The Transportation Problem

Operations research (OR) is a branch of applied mathematics that deals with researching the operations of organizations, such as businesses and industries, with the goal of helping them operate more efficiently. It is also known as management science and is closely related to the field of engineering called industrial engineering. Operations research began in earnest during World War II as an attempt to help the military allocate and transport resources more efficiently.

Operations research involves topics such as queuing theory, inventory theory, and simulations. Solving LPs is also a big part of OR. In OR, LPs are categorized according to the form of the model and the different categories are named according to the prototypical example of the form. In this and the next section we consider two very common types of LPs; the *transportation problem* and the *assignment problem*.

Example 7.3.1   Delivering Bread
The Better Bread Company has three bakeries located in the midwest United States near wheat-growing areas and four distribution warehouses scattered across the United States. Management is studying ways to reduce shipping costs. Table 7.5 summarizes the weekly output of each bakery and the weekly allocation of each warehouse (in units of truckloads), along with the estimated shipping costs for each bakery–warehouse combination. Determine how many truckloads of bread should be assigned to each bakery–warehouse combination to minimize total cost.

Warehouse

Bakery	1	2	3	4	Output
1	119	253	321	402	21
2	205	198	348	365	18
3	432	351	195	248	18
Allocation	15	10	12	20	

Table 7.5

To form the model, let x_{ij} $(i = 1, 2, 3; j = 1, 2, 3, 4)$ represent the number of truckloads to be shipped from bakery i to warehouse j. This may be formulated as a linear program as

Minimize

$C = 119\, x_{11} + 253x_{12} + \cdots + 248x_{34}$

Subject to

$$x_{11} + x_{12} + x_{13} + x_{14} = 21$$
$$x_{21} + x_{22} + x_{23} + x_{24} = 18$$
$$x_{31} + x_{32} + x_{33} + x_{34} = 18$$
$$x_{11} + x_{21} + x_{31} = 15$$
$$x_{12} + x_{22} + x_{32} = 10$$
$$x_{13} + x_{23} + x_{33} = 12$$
$$x_{14} + x_{24} + x_{34} = 20$$
$$x_{ij} \geq 0 \text{ for all } i = 1, 2, 3 \text{ and } j = 1, 2, 3, 4.$$

The first constraint says that the total number of trucks coming out of bakery 1 must equal its total output. The next two constraints have similar meanings for bakeries 2 and 3. The fourth constraint says that the total number of trucks going to warehouse 1 must equal its total allocation. The next three constraints have similar meanings for warehouses 2, 3, and 4.

Notice the special pattern of coefficients in the constraints. It is this pattern that sets the transportation problem apart from others. Any problem that can be completely described by a parameter table like that in Table 7.5 will have this pattern and is thus called a "transportation problem" regardless of whether it has anything to do with transportation.

Also notice that the total output from the bakeries equals the total allocation of the warehouses. In more general terminology we say the total *supply* from the *sources* equals the total *demand* from the *destinations*. This is a necessary condition for there to be a solution to this problem.

To solve this problem in Excel, format a blank worksheet as in Figure 7.5 and format the Solver window as in Figure 7.6.

	B	C	D	E	F	G
1	Data					
2			Warehouse			
3	Bakery	1	2	3	4	Output
4	1	119	253	321	402	21
5	2	205	198	348	365	18
6	3	432	351	195	248	18
7	Allocation	15	10	12	20	
8						
9	Solution					
10			Warehouse			
11	Bakery	1	2	3	4	Total
12	1	15	0	6	0	=SUM(C12:F12)
13	2	0	10	0	8	=SUM(C13:F13)
14	3	0	0	6	12	=SUM(C14:F14)
15	Total	=SUM(C12:C14)	=SUM(D12:D14)	=SUM(E12:E14)	=SUM(F12:F14)	
16						
17			Total Cost			
18			=SUMPRODUCT(C4:F6,C12:F14)			

Figure 7.5

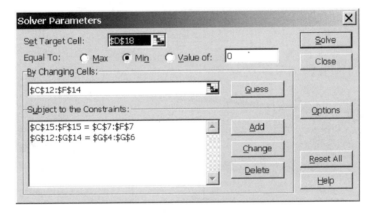

Figure 7.6

The solution is shown in Figure 7.7. It says that bakery 1 should ship 15 truckloads to warehouse 1, and 3 to warehouse 3. Bakery 2 should ship 10 truckloads to warehouse 2, and 8 to warehouse 4. Bakery 3 should ship 6 truckloads to warehouse 3, and 12 to warehouse 4.

Example 7.3.2   Different Allocations
Consider the same problem as in Example 7.3.1, but with slightly different warehouse allocation amounts as given in Table 7.6.

	B	C	D	E	F	G
10			Warehouse			
11	**Bakery**	1	2	3	4	**Total**
12	1	15	0	6	0	21
13	2	0	10	0	8	18
14	3	0	0	6	12	18
15	**Total**	15	10	12	20	

Figure 7.7

Bakery	Warehouse				Output
	1	2	3	4	
1	119	253	321	402	21
2	205	198	348	365	18
3	432	351	195	248	18
Allocation	13	10	12	19	

Table 7.6

Notice that the total output is 57 and the total allocation is 54. Therefore, we cannot require all bakeries to produce all of their potential output since there isn't enough warehouse space to hold it all. We could model this problem as an LP using inequality constraints as

Minimize

$C = 119\,x_{11} + 253x_{12} + \ldots + 248x_{34}$

Subject to

$$x_{11} + x_{12} + x_{13} + x_{14} \leq 21$$
$$x_{21} + x_{22} + x_{23} + x_{24} \leq 18$$
$$x_{31} + x_{32} + x_{33} + x_{34} \leq 18$$
$$x_{11} + x_{21} + x_{31} = 13$$
$$x_{12} + x_{22} + x_{32} = 10$$
$$x_{13} + x_{23} + x_{33} = 12$$
$$x_{14} + x_{24} + x_{34} = 19$$
$$x_{ij} \geq 0 \text{ for all } i = 1, 2, 3 \text{ and } j = 1, 2, 3, 4.$$

This LP could be solved with Solver using the simplex method. However, the formulation above is not a transportation problem since there are some inequality constraints instead of all equality constraints. This may not seem like a big deal. However, there is a special version of the simplex method, called the *transportation simplex method*, which is much more efficient at solving transportation problems than the generic simplex method. For large problems with thousands of decision variables (as occur in real applications), the transportation simplex method could save a great deal of computation time over the simplex method. Therefore, if a problem can be modeled as a transportation problem, it is preferable to do so.

To model this problem as a transportation problem, we introduce a "dummy" warehouse that will be allocated the excess three trucks of bread and given shipping costs of 0. The parameter table including this dummy warehouse is shown in Table 7.7.

Warehouse

Bakery	1	2	3	4	Dummy	Output
1	119	253	321	402	0	21
2	205	198	348	365	0	18
3	432	351	195	248	0	18
Allocation	13	10	12	19	3	

Table 7.7

Notice that in this version the total allocation equals the total output, so this version is a true transportation problem. Note that this problem has three additional decision variables. To solve this problem, we can simply modify the worksheet **Bread** to include this dummy warehouse. (**Note:** Be sure to modify the formulas for the Totals and the Total Cost appropriately.) The solution is shown in Figure 7.8.

	B	C	D	E	F	G	H
10			Warehouse				
11	**Bakery**	1	2	3	4	Dummy	Total
12	1	13	0	5	0	3	21
13	2	0	10	0	8	0	18
14	3	0	0	7	11	0	18
15	Total	13	10	12	19	3	

Figure 7.8

The solution says to ship 3 truckloads from bakery 1 to the dummy warehouse. This means in practice that bakery 1 should produce only 18 truckloads, 13 of which go to warehouse 1, and 5 of which go to warehouse 3.

Example 7.3.3   A New Requirement

Consider the same scenario as in Example 7.3.2, except management is now requiring that bakery 1 produce exactly 21 truckloads. We could model this as before, except with the added constraint that the number of truckloads going from bakery 1 to the dummy warehouse is exactly 0, but then we wouldn't have a transportation problem.

Instead, we will use the *"big-M" method*. We model the problem exactly as in Example 7.3.2, except we will assign a large cost, M (in this case M = 1000), to a shipment from bakery 1 to the dummy warehouse. The parameter table is shown in Table 7.8.

Warehouse

Bakery	1	2	3	4	Dummy	Output
1	119	253	321	402	1000	21
2	205	198	348	365	0	18
3	432	351	195	248	0	18
Allocation	13	10	12	19	3	

Table 7.8

Solving this requires us to change only this one cost value in the modified worksheet **Bread**. The solution is shown in Figure 7.9. Notice that no truckload is shipped from bakery 1 to the dummy warehouse, as required. In this solution, bakery 2 should produce only 15 truckloads.

	B	C	D	E	F	G	H
10			Warehouse				
11	**Bakery**	1	2	3	4	**Dummy**	**Total**
12	1	13	0	8	0	0	21
13	2	0	10	0	5	3	18
14	3	0	0	4	14	0	18
15	**Total**	13	10	12	19	3	

Figure 7.9

Exercises

Directions: Formulate each problem below as a transportation problem by constructing an appropriate parameters table as in the "Data" section of Figure 7.5 and solve it with Solver. The parameter tables should meet the following requirements:

1. The "sources" are listed along the left-hand side.

2. The "destinations" are listed along the top.

3. The "supply" from each source is listed along the right-hand side.

4. The "demand" from each destination is listed along the bottom.

5. The total output equals the total demand.

7.3.1 The manager of the produce department at a local supermarket buys her strawberries from two local suppliers, Sunnyside Farms and Green Valley Farms. The manger needs 3 cases of strawberries today and an additional 7 cases tomorrow. Sunnyside Farms can sell a maximum of 6 cases total at a price of $7.25 per case today and $6.35 per case tomorrow. Green Valley Farms can sell a maximum of 4 cases total at a price of $5.75 per case today and $5.25 per case tomorrow. How should the manager make her purchases to minimize the total cost while still meeting her daily requirements? (**Hint:** Let the sources be Sunnyside Farms and Green Valley Farms and the destinations be today and tomorrow. The costs would then be Sunnyside's price to sell today, Green Valley's price to sell tomorrow, and so on.)

7.3.2 The Great Openings Company, which manufactures doors and windows, is reassigning the production of three of its products to five of its plants. The costs to manufacture one unit of each product in each plant, the capacity of each plant, and the anticipated demand for each product are shown in Table 7.9.

Product

Plant	1	2	3	Capacity
1	25	50	40	300
2	32	48	36	450
3	35	46	42	625
4	18	38	31	700
5	36	45	30	725
Demand	900	1100	600	

Table 7.9

1. Management needs to know how to assign the products to the plants to minimize total manufacturing cost.

2. Suppose plants 4 and 5 cannot manufacture product 2 and plant 1 must manufacture exactly 300 units. Find the new optimal assignment.

7.3.3 The Better Bread Company also produces fruit cakes, most of which are sold during the holiday season. In September and October, the company ramps up its fruit cake production capacity, but then starts to cut back in November and December. The anticipated monthly demand, maximum production capacity, and production costs are shown in Table 7.10 (one unit is 1000 fruit cakes and costs are in thousands of dollars). Fruit cakes can be produced in one month, stored, and then sold in another month, but there is a 0.0015 monthly cost to warehouse each unit. How should the monthly production be scheduled to minimize total cost while still meeting the anticipated demand? (**Hint:** Let the sources be the months produced and the destinations be the months sold. The cost c_{ij} would then be the total cost for a unit of fruit cakes that is produced in month i and sold in month j. For instance, $c_{11} = 1.18$ and $c_{14} = 1.1845$. Is it possible to produce a unit of fruit cakes in month 3 and sell it in month 1? What does this mean about c_{31}?)

Month	Anticipated Demand	Production Capacity	Unit Production Cost	Unit Storage Cost
Sept	5	35	1.18	0.0015
Oct	10	50	1.21	0.0015
Nov	40	40	1.19	0.0015
Dec	60	10	1.2	

Table 7.10

7.3.4 (This is not a transportation problem, so no parameters table is needed, but the decision variables do have similar meanings as in a transportation problem.) The Nutty Goodness Company is considering selling three new mixtures of peanuts, walnuts, and cashews. It has 1000 lb of peanuts, 800 lb of walnuts, and 700 lb of cashews available for the first batch. Each mixture has a unique set of specifications as to the percentage of each type of nut as shown in Table 7.11.

Mixture	Specifications	Selling Price per Pound
Walnut Lover's	At least 35% walnuts At most 25% cashews No restriction on peanuts	$1.85
Cashew Lover's	At most 45% peanuts At least 45% cashews No restriction on walnuts	$1.99
Premium	At most 10% peanuts Between 50% and 65% walnuts At least 15% cashews	$2.75

Table 7.11

If peanuts cost $0.75 per lb, walnuts cost $1.05 per lb, and cashews cost $1.75 per lb, determine how much of each mixture the company should make (and the amount of each type of nut in each mixture) to maximize total profit. (**Hint:** Let x_{ij} = amount of nut i in mixture j. In your worksheet, set up different cells to calculate the total amount of each mixture, the total amount of each nut in each mixture, and the percentage of each type of nut in each mixture. Also calculate total cost and total revenue. Remember, profit = revenue − cost.)

7.4   The Assignment Problem and Binary Constraints

The *assignment problem* (AP) is a special type of linear programming problem where we generically say that *workers* are being assigned to perform *jobs*. The simplest example is when employees are being assigned different types of jobs to perform. However, the workers may not always be people and the jobs may not always be jobs to perform. In Example 7.4.1 below, the workers are trucks and the jobs are packages that need to be delivered.

In an AP, the decisions to be made are which workers should perform which jobs so as to minimize some cost function. A problem is called an AP if and only if it satisfies the following assumptions (Hillier and Lieberman 2001, 382):

1. The number of workers, n, equals the number of jobs.
2. Each worker is to be assigned exactly one job.
3. Each job is to be performed by exactly one worker.
4. There is a cost c_{ij} associated with assigning worker i to job j.
5. The objective is to determine how all n assignments should be made to minimize the total cost.

The "cost" in the fourth assumption may be a literal dollar amount associated with assigning worker i to job j, or it may be some other value associated with that assignment (e.g., time), the total of which we want to minimize. The decision variables, x_{ij} for i, $j = 1$, $2, \ldots, n$, are binary (i.e., they equal 0 or 1) with

$$x_{ij} = \begin{cases} 1 & \text{if worker } i \text{ performs job } j \\ 0 & \text{if not.} \end{cases}$$

Assignment problems can be solved extremely efficiently by a special form of the transportation simplex method (which is in turn a special form of the simplex method).

Example 7.4.1   On-Time Delivery Company

A scheduler at the On-Time Delivery Company has three packages that need to be delivered and four available trucks of different types. Based on the locations of the packages and the locations of the trucks, the scheduler has determined a cost for each truck to deliver each package as shown in Table 7.12 (truck 2 cannot deliver package 2).

The goal is to determine how to assign trucks to packages to minimize the total delivery costs. Note that this is technically not an AP since there are more workers than jobs, so we will add a dummy package. Each truck will "supply" one delivery and each package will "demand" one delivery. Also, we will use the big-M method to model the fact that truck 2 cannot deliver package 2. The resulting parameters table is shown in Table 7.13.

Package

Truck	1	2	3
1	12	20	13
2	16	–	18
3	8	6	9
4	14	10	8

Table 7.12

Package

Truck	1	2	3	Dummy	Supply
1	12	20	13	0	1
2	16	100	18	0	1
3	8	6	9	0	1
4	14	10	8	0	1
Demand	1	1	1	1	

Table 7.13

Notice that this is an appropriate parameters table for a transportation problem. So an AP is indeed a transportation problem. However, an AP has the additional feature that the supplies and the demands are all 1. This special feature can be exploited to find an algorithm that solves these problems extremely efficiently. Note that because the AP is a transportation problem, the supplies and demands are all integers, and the problem has a special structure, the optimal solution will be binary (i.e., we don't have to add the constraints that the decision variables are binary; we get this for free).

Solving this problem in the same way we did in Section 7.3 yields the solution shown in Figure 7.10. This shows that truck 1 should deliver package 1, truck 3 should deliver package 2, truck 4 should deliver package 3, and truck 2 should not deliver any package.

	B	C	D	E	F	G
11			Package			
12	Truck	1	2	3	Dummy	Sum
13	1	1	0	0	0	1
14	2	0	0	0	1	1
15	3	0	1	0	0	1
16	4	0	0	1	0	1
17	Sum	1	1	1	1	
18			Total Cost =	26		

Figure 7.10

Example 7.4.2   Ace Manufacturing

Ace Manufacturing Company is going to start manufacturing four new products in three existing plants that have excess capacity. Table 7.14 shows the daily cost of producing one unit of each product in each plant (note plant 2 cannot produce product 3), the available daily capacity of each plant, and the expected daily demand for each product.

Management wants each product to be produced in exactly one plant (e.g., we cannot produce 12 units of product 1 in plant 1 and 13 units in plant 3). How should they schedule production to minimize the total production cost while still meeting demand?

	Product				
Plant	1	2	3	4	Capacity
1	35	32	30	40	78
2	35	25	–	31	78
3	33	37	36	30	45
Demand	25	35	32	43	

Table 7.14

We model this problem as an AP. Note that plants 1 and 2 have enough capacity to produce up to two products. Plant 3 can produce at most one product. So we split plants 1 and 2 into two plants each (call them plants 1a, 1b, 2a, and 2b, respectively), thus creating five plants each of which has a "supply" of 1. These five plants are the workers. Therefore, we need five products (or jobs), so we create a dummy product. The "demand" for each product is 1, and the cost of assigning a product to a plant is the total daily cost for producing all of that product in that plant. These costs are equal to the unit costs times the daily demands as given in Table 7.14. Also, we use the Big-M method to model the fact that plant 2 cannot manufacture product 3. The parameters table for this formulation is shown in Table 7.15.

	Product					
Plant	1	2	3	4	Dummy	Supply
1a	875	1120	960	1720	0	1
1b	875	1120	960	1720	0	1
2a	875	875	20,000	1333	0	1
2b	875	875	20,000	1333	0	1
3	825	1295	1152	1290	0	1
Demand	1	1	1	1	1	

Table 7.15

Solving this problem yields the solution shown in Figure 7.11, which says to produce product 3 in plant 1, products 2 and 4 in plant 2, and product 1 in plant 3.

	B	C	D	E	F	G	H
10				Product			
11	Plant	1	2	3	4	Dummy	Total
12	1a	0	0	1	0	0	1
13	1b	0	0	0	0	1	1
14	2a	0	0	0	1	0	1
15	2b	0	1	0	0	0	1
16	3	1	0	0	0	0	1
17	Total	1	1	1	1	1	
18			Total Cost =	3993			

Figure 7.11

As we have seen in the AP, binary decision variables can be used to represent yes–no decisions. Many other types of scenarios involve yes–no decisions and can thus be modeled with binary decision variables. The next example illustrates one such scenario.

Example 7.4.3 Home Improvement Decisions
Consider the following scenario:

> Steve and Laura are trying to sell their house, which has two bedrooms and two bathrooms. To increase the house's value, they want to remodel one or more rooms. They have estimated the costs of remodeling each room and their real estate agent has estimated the increase in the house's value if each room was remodeled as shown in Table 7.16 (where costs and increases in values are given in thousands of dollars). They have only \$10,000 to spend remodeling, and they have decided that they cannot do both bathroom 2 and bedroom 2. They will only do bathroom 2 if they also do bathroom 1. Also, they will only do bedroom 2 if they also do bedroom 1. Which rooms should Steve and Laura remodel to maximize the total increase in their house's value?

Room	Decision Variable	Remodeling Cost	Increase in House Value
Bathroom 1	x_1	6	9
Bedroom 1	x_2	3	5
Bathroom 2	x_3	5	6
Bedroom 2	x_4	2	4

Table 7.16

There are four decisions to make in this problem: Do they remodel Bathroom 1? Do they remodel Bedroom 1? and so on. Each one of the four decision variables will equal 1 if the associated decision is yes and 0 if the decision is no.

Our objective is then to maximize

$$Z = 9x_1 + 5x_2 + 6x_3 + 4x_4.$$

The fact that they have only $10,000 to spend means that

$$6x_1 + 3x_2 + 5x_3 + 2x_4 \leq 10.$$

Since they cannot do both bathroom 2 and bedroom 2, we cannot have $x_3 = x_4 = 1$. So in terms of inequalities, we have

$$x_3 + x_4 \leq 1.$$

Since they will only do bathroom 2 if they also do bathroom 1, we can only have $x_1 = 1$ and $x_3 = 1$, or $x_1 = 1$ and $x_3 = 0$, or $x_1 = 0$ and $x_3 = 0$. In terms of inequalities, we have

$$x_3 \leq x_1 \implies -x_1 + x_3 \leq 0.$$

Likewise, since they will only do bedroom 2 if they also do bedroom 1, we have

$$x_4 \leq x_2 \implies -x_2 + x_4 \leq 0.$$

Putting it all together we get the program

$$
\begin{aligned}
\textbf{Maximize } Z = \;& 9x_1 + 5x_2 + 6x_3 + 4x_4 \\
\textbf{Subject to} \quad & 6x_1 + 3x_2 + 5x_3 + 2x_4 \leq 10 \\
& \qquad\qquad\; x_3 + x_4 \leq 1 \\
& -x_1 \qquad\; + x_3 \qquad\;\; \leq 0 \\
& \qquad -x_2 \qquad\; + x_4 \leq 0 \\
& x_1,\, x_2,\, x_3,\, x_4 \text{ are binary.}
\end{aligned}
$$

Notice that this program does not fit the form of a transportation problem, so we do need the additional binary constraints. To solve this in Excel, rename a blank worksheet "**Improvement**" and format it as in Figure 7.12.

	A	B	C	D	E	F	G	H	I
1		Bath 1	Bed 1	Bath 2	Bed 2				
2	Variable	x_1	x_2	x_3	x_4				
3	Values	0	0	0	0				
4	Z =	9	5	6	4	=	=SUMPRODUCT($B$3:$E$3,B4:E4)		RHS
5	Constraint 1	6	3	5	2	=	=SUMPRODUCT($B$3:$E$3,B5:E5)	≤	10
6	Constraint 2			1	1	=	=SUMPRODUCT($B$3:$E$3,B6:E6)	≤	1
7	Constraint 3	-1		1		=	=SUMPRODUCT($B$3:$E$3,B7:E7)	≤	0
8	Constraint 4		-1		1	=	=SUMPRODUCT($B$3:$E$3,B8:E8)	≤	0

Figure 7.12

Format the Solver window as in Figure 7.13.

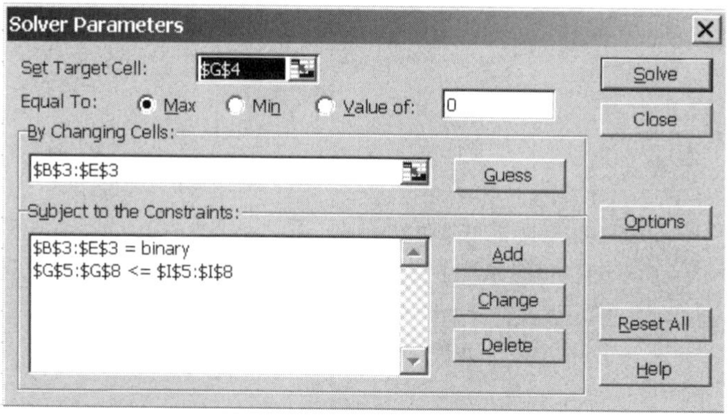

Figure 7.13

The results are shown in Figure 7.14. They indicate that Steve and Laura should remodel bedroom 1 and bathroom 1, which will increase their house's value by $14,000.

	A	B	C	D	E	F	G	H	I
1		Bath 1	Bed 1	Bath 2	Bed 2				
2	Variable	x_1	x_2	x_3	x_4				
3	Values	1	1	0	0				
4	Z =	9	5	6	4	=	14		RHS
5	Constraint 1	6	3	5	2	=	9	≤	10
6	Constraint 2			1	1	=	0	≤	1
7	Constraint 3	-1		1		=	-1	≤	0
8	Constraint 4		-1		1	=	-1	≤	0

Figure 7.14

Exercises

Directions: Solve Exercises 7.4.1 and 7.4.2 below by formulating each as an assignment problem and constructing an appropriate parameters table that meets the following requirements:

1. The "workers" are listed along the left-hand side.
2. The "jobs" are listed along the top.
3. The number of "workers" equals the number of "jobs."
4. The "supply" of each worker is 1 and the "demand" of each job is 1.

7.4.1 The final round of a mathematics competition consists of four tests: arithmetic, algebra, calculus, and geometry. Each team is supposed to assign exactly one member to take each test. The team consisting of Brian, Ed, Larry, John, and Bruce has personal-best scores on each test as shown in Table 7.17. How should they assign themselves to the tests so that no one takes more than one test and the sum of the corresponding scores is maximized?

Test	Brian	Ed	Larry	John	Bruce
Arithmetic	99.1	96.3	97.6	98.9	98.5
Algebra	99.3	98.9	98.2	99.2	98.8
Calculus	94.6	98.3	97.4	97.2	98.2
Geometry	95.3	98.5	98.6	98.5	94.2

Table 7.17

7.4.2 A farmer, Stan, is going to harvest wheat on three different fields. He can haul the grain to two different elevators: the Farmers United elevator, which pays \$4.50 per bushel and can accept only 1800 bushels of wheat; and the Sunflower elevator, which pay \$4.30 per bushel and can accept only 1000 bushels. He predicts that fields 1, 2, and 3 will produce 1000, 500, and 1000 bushels, respectively. His grain trucks can haul 500 bushels each. Assume that only full trucks will be used to haul the wheat to the elevators. The price to haul one bushel from each field to each elevator is shown in Table 7.18.

	Field		
Elevator	1	2	3
Farmers United	0.13	0.13	0.15
Sunflower	0.16	0.13	0.17

Table 7.18

Stan needs to determine how much to haul from each field to each elevator to maximize the total profit. [**Hint:** The profit for a bushel is the price paid at the elevator minus the hauling cost. Consider having five jobs (or truckloads of wheat) and five workers. The "cost" for each assignment is the profit for each truckload.]

7.4.3 Suppose Steve and Laura are going to start their remodeling project. They have decided to do the drywall, painting, trim work, and finish plumbing themselves. They want to divide these four tasks between them so that each has exactly two tasks, but the total time they take is kept to a minimum. Each person has estimated the amount of time he or she will take for each task as shown in Table 7.19.

	Drywall	Painting	Trim Work	Finish Plumbing
Steve	10.0	8.5	9.5	3.0
Laura	9.5	6.0	9.0	4.0

Table 7.19

Laura refuses to do both the painting and the drywall. Steve won't do the finishing plumbing unless Laura does the trim work. How should they divide the tasks between them? Formulate this problem using binary decision variables and solve it with Solver. (**Hint:** Consider eight decision variables, one for each person–task combination. You will need two constraints to make sure that each person does exactly two tasks. Four are needed to make sure that each task gets assigned. Two more are necessary for the additional requirements.)

7.5   Solving Linear Programs

In Section 7.2 we looked at the following linear problem:

The manager of a produce department at a neighborhood supermarket is making fruit baskets for the busy holiday season. He sells two sizes of baskets: small and large. He has only 200 apples and 100 oranges remaining and is trying to decide how many of each size of basket he should make. Each small basket returns a profit of $3 and requires 3 apples and 1 orange while each large basket returns a profit of $4 and requires 2 apples and 2 oranges. Assuming he will sell all that he makes, how many of each size basket should he make to maximize his profit?

We modeled this problem with the linear program:

$$\text{Maximize } P = 3s + 4l$$
$$\text{Subject to} \quad 3s + 2l \leq 200$$
$$1s + 2l \leq 100 \tag{7.2}$$
$$s, l \geq 0$$

where s = the number of small baskets to produce and l = the number of large baskets to produce. Solver gave an optimal solution of 50 small and 25 large baskets, which gives a maximum profit of $250. In this section we look at a graphical and an algebraic way to solve this program. In the next section we examine how the simplex method works.

Example 7.5.1   Graphical Solution

To examine a graphical solution to the model (7.2), open the worksheet "**Graphical Solution**" in the workbook **Linear Programming**, found on the website for this book, and follow these steps:

1. Enter the names of the decision variables, the objective functions, and the first two constraints as shown in Figure 7.15 (don't worry about the nonnegativity constraints).

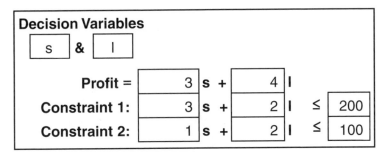

Figure 7.15

2. The resulting graph is shown in Figure 7.16 (s is on the horizontal axis and l is on the vertical axis). The graph contains the lines $3s + 2l = 200$ and $s + 2l = 100$, called the *constraint lines* (the nonnegativity constraints mean that the s- and l-axes are also constraint lines). The area below the first line is the set of all points (i.e., combinations of numbers of small and large baskets) that satisfy the first constraint. Points below the second line satisfy the second constraint. Points below *both* lines satisfy *both* constraints and constitute what is called the *feasible region*. These points are all the feasible solutions.

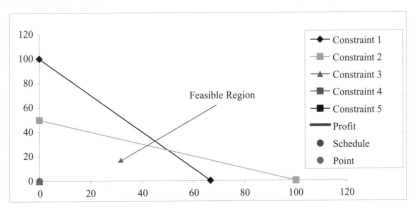

Figure 7.16

3. Move the slider under **Desired Profit** to the right until the desired profit is 170.00. Move the slider under **Desired Schedule** to the right until $s = 22.0$ and $l = 26.0$.

The resulting graph is shown in Figure 7.17. The thick black line labeled "Profit," called a *level curve*, is the set of all points that will give a profit of $170.00. Notice that this line intersects the feasible region. This means that it is possible to produce a combination of small and large baskets that gives a profit of $170.00. Specifically, the point $s = 22.0$ and $l = 26.0$ is on this level curve and in the feasible region. This means it is possible to produce 22 small baskets, 26 large baskets, with a profit of $170.00.

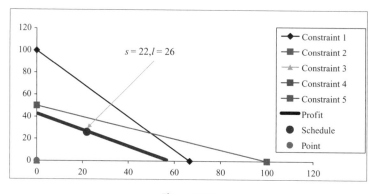

Figure 7.17

4. Continue to move the slider under **Desired Profit** to the right to increase the desired profit. Notice that for profits above 250.00, the level curve does not intersect the feasible region. This means that it is *not* possible to profit more than $250.00. This is our maximum profit. Move the slider until the desired profit is exactly 250.00 and note that the level curve intersects the feasible region at only one point. Using the slider under **Desired Schedule** we see that the coordinates of this point are $s = 50$ and $l = 25$, as shown in Figure 7.18. This is our optimal solution, which is exactly the same as that found by Solver.

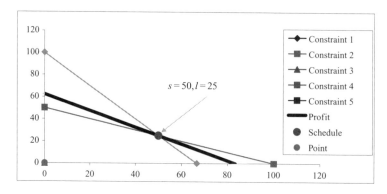

Figure 7.18

The feasible region in this, and all linear programs, forms what is called a *convex set*, which means that if any two points in the set are joined by a line segment, the segment lies entirely within the set (i.e., it never "leaves" the set). A *corner point* of the feasible region is a point of intersection of two or more constraint lines. Note that the optimal solution to this problem occurs at a corner point. For this reason, the solution is often called a *corner-point solution*. The feasible, but nonoptimal, solution $s = 22.0$ and $l = 27.0$ is called an *interior-point solution*.

These observations are generalized in Theorem 7.5.1.

Theorem 7.5.1 *If the convex feasible region of a linear program is nonempty and bounded, then the maximum and minimum values of the objective function will occur at corner points of the region. If the feasible region is unbounded, then the objective function may not attain maximum or minimum values.*

Theorem 7.5.1 is important because it tells us that to find the optimal solution to a linear program, we only need to consider the corner points of the feasible region. This observation forms the basis of the algebraic solution and the simplex method.

Example 7.5.2   Algebraic Solution

The fact that the optimal solution to a linear programming problem occurs at a corner point of the feasible region suggests that we simply need to find all the corner points and pick the best one. To illustrate this idea, again consider the linear program (7.2):

$$\text{Maximize } P = 3s + 4l$$
$$\text{Subject to} \quad 3s + 2l \le 200$$
$$1s + 2l \le 100$$
$$s, l \ge 0.$$

To find the corner points, we first find all the intersection points of the constraint lines. To do this, we translate the first two inequality constraints into equality constraints by introducing the "slack" variables y_1 and y_2:

$$\text{Maximize } P = 3s + 4l$$
$$\text{Subject to} \quad 3s + 2l + y_1 \qquad = 200$$
$$1s + 2l \qquad + y_2 = 100 \tag{7.3}$$
$$s, l, y_1, y_2 \ge 0.$$

The variable y_1 is called the slack variable because it takes up the difference (i.e., the "slack") between the quantity $3s + 2l$ and the number 200. Since $3s + 2l$ must be less than or equal to 200, y_1 must be nonnegative. A similar explanation applies to y_2.

Now to calculate the points of intersection, consider the graph of the constraints shown in Figure 7.16. For any point on the constraint 1 line,

$$3s + 2l = 200,$$

so $y_1 = 0$. For any point on the constraint 2 line,

$$1s + 2l = 100,$$

so $y_2 = 0$. Therefore, any point at the intersection of these two constraint lines will be characterized by $y_1 = y_2 = 0$. Thus, to find this point of intersection, we could set $y_1 = y_2 = 0$ in (7.3) and solve the resulting system of linear equations

$$3s + 2l = 200$$
$$1s + 2l = 100$$

for s and l, yielding $s = 50$, $l = 25$, respectively. This is one corner point.

The s-axis is the constraint line for the constraint $l \geq 0$. Obviously for any point on this line, $l = 0$. Therefore, any point at the intersection of constraint 1 and the constraint $l \geq 0$ is characterized by $y_1 = l = 0$. To find the point of intersection of these two constraints, we could set $y_1 = l = 0$ in (7.3) and solve the resulting system

$$3s \qquad = 200$$
$$1s + y_2 = 100$$

for s and y_2. This yields $s = 200/3$, $y_2 = 100/3$. Thus another corner point is $s = 200/3$, $l = 0$.

In general, to find the points of intersections of the constraint lines we need to set two variables in (7.3) equal to 0 and solve for the remaining variables. The number of ways we could choose which variables to set equal to 0 (i.e., the number of points of intersection) is given by

$$\binom{4}{2} = \frac{4!}{2!\,(4-2)!} = 6.$$

All of these different combinations are shown in Table 7.20. Note that not all of these are feasible because some don't satisfy the nonnegativity constraints in (7.3). We see that we have four feasible corner-point solutions, and the optimal one is $s = 50$, $l = 25$ (i.e., 50 small and 25 large baskets, respectively) with a profit of $250. This is the exact same solution we got graphically and with Solver.

Point	s	l	y_1	y_2	Feasible?	Profit
1	0	0	200	100	Yes	0
2	0	100	0	-100	No	–
3	0	50	100	0	Yes	200
4	200/3	0	0	100/3	Yes	200
5	100	0	-100	0	No	–
6	50	25	0	0	Yes	250

Table 7.20

This basic idea seems simple enough, but there is a major problem with computational efficiency. Suppose we have a linear program with m decision variables and n inequality constraints. First we convert the constraints into equalities by introducing n slack variables, similar to above. This gives a total of $m + n$ variables. We find the points of intersection of the constraint lines by setting m variables equal to 0. This gives a system of n equations with n variables, which can be solved, in principle, by matrix techniques.

The number of ways we could choose these m variables to set equal to 0 (i.e., the number of points of intersection) is then given by

$$\binom{m+n}{m} = \frac{(m+n)!}{m!\,((m+n)-m)!} = \frac{(m+n)!}{m!\,n!}.$$

For a problem with 25 decision variables and 50 inequality constraints (which is relatively small for a real application), the number of points of intersection is

$$\binom{75}{25} = \frac{75!}{25!\,50!} = 5.25 \times 10^{19}.$$

This is far too many points to check for any computer. So we want to find an algorithm that does not require finding and checking all points of intersection. One such algorithm is the simplex method, which is discussed in the next section.

Exercises

7.5.1 Suppose the produce manager starts making 50 small and 25 large baskets but then discovers that 2 apples and 5 oranges are rotten. Is it still possible to make 50 small and 25 large baskets total? Why or why not? How many baskets should he make now to maximize his profit?

7.5.2 Suppose the produce manager adds the constraint that he cannot make more than 40 small baskets (assume he still has 200 apples and 100 oranges).

1. Graphically estimate the new optimal solution.

2. Suppose he limits himself to 30 small baskets. What is the new optimal solution?

3. As the number of allowed small baskets decreases, what happens to the optimal solution graphically?

4. Suppose we add the generic constraint $s \leq s_0$ where $0 \leq s_0 \leq 50$. Find a formula for the optimal solution (remember, s and l must be integers).

7.5.3 Consider the linear program:

$$\begin{aligned}
\text{Minimize } C = {} & x + 2y \\
\text{Subject to} \quad & x + y \geq 6 \\
& 3x + y \geq 9 \\
& x, \, y \geq 0.
\end{aligned}$$

1. Graphically solve this program (note that the feasible region is *above* the constraint lines).

2. Find a value of the coefficient of x in the objective function so that the optimal solution is $x = 1.5$, $y = 4.5$ with $C = 15.00$.

7.5.4 Solve the linear program

$$\begin{aligned}
\text{Maximize } P = {} & 20x + 32y \\
\text{Subject to} \quad & 6x + y \leq 6 \\
& 3x + 2y \leq 9 \\
& x, \, y \geq 0
\end{aligned}$$

algebraically by enumerating all the possible corner points, determining which are feasible, and choosing the best one as done in Table 7.20. Verify your solution using Solver.

7.6 The Simplex Method

Consider the linear program (7.2) (rewritten with decision variables x_1 and x_2 and objective function value z):

$$\begin{aligned}
\text{Maximize } z = {} & 3x_1 + 4x_2 \\
\text{Subject to} \quad & 3x_1 + 2x_2 \leq 200 \\
& x_1 + 2x_2 \leq 100 \\
& x_1, \, x_2 \geq 0.
\end{aligned}$$

We solved this program with Solver, graphically, and algebraically and got an optimal solution of $x_1 = 50$, $x_2 = 25$ with a profit of 250.

In this section we illustrate some of the basic ideas behind the simplex method by using it to solve this problem. Note that the version of the simplex method we discuss here applies only to maximization linear programs with inequality constraints of the form $\leq$. To apply this version to a problem of another type would require reformatting the problem into this basic form.

Graphical Interpretation

Consider the graph of the feasible region of this problem in Figure 7.16. We know that the optimal solution will lie at one of the corner points of this feasible region. Graphically the simplex method works by moving from one corner point to an adjacent corner point on the border of the feasible region (the border is called a *simplex*).

1. Start at a corner point (typically we start at $(0,0)$).
2. Move either up or to the right. This corresponds to increasing the value of x_1 or x_2.
3. To determine which direction to move, we look at the objective function $z = 3x_1 + 4x_2$. We want to maximize its value, so increasing x_2 will increase its value faster than increasing x_1. So, we move up.
4. To determine how far up we can go, we look at the constraints

$$3x_1 + 2x_2 \leq 200$$
$$1x_1 + 2x_2 \leq 100.$$

Since the decision variables are nonnegative, these constraints tell us that to remain feasible, at the very least we must have

$$2x_2 \leq 200 \quad \Rightarrow \quad x_2 \leq 100$$
$$\text{and}$$
$$2x_2 \leq 100 \quad \Rightarrow \quad x_2 \leq 50.$$

These four steps move us from the first corner point to the next one. The basic idea behind the next move is the same, but performing the calculations requires us to rewrite the program. This is done most easily with matrices.

Tableau Form

To rewrite the program using matrices, we convert it into "tableau" form by adding slack variables, converting the constraints to equalities, and rewriting the objective function and adding it to the constraints:

$$\text{Maximize} \quad z$$
$$\begin{aligned} \text{Subject to} \quad 3x_1 + 2x_2 + y_1 \qquad\quad &= 200 \\ x_1 + 2x_2 \quad + y_2 \quad &= 100 \\ -3x_1 - 4x_2 \qquad\qquad + z &= 0 \\ x_1, \, x_2, \, y_1, \, y_2 \geq 0. \end{aligned}$$

Now we can use this form of the problem to implement the simplex method. Use the worksheet "**2 Variable Simplex Method**" in the workbook **Linear Programming**, found on the website for this book, to perform these steps:

Step 1: Form an initial **tableau** as in Figure 7.19 (RHS stands for right-hand side). This tableau is really nothing more than a matrix of the coefficients in the tableau form of the program.

Tableau 0

Basic	x_1	x_2	y_1	y_2	z	RHS	Ratio
	3	2	1	0	0	200	
	1	2	0	1	0	100	
	-3	-4	0	0	1	0	

Figure 7.19

In the algebraic solution to this problem, we set two variables equal to 0 and solved for the others. Here, we take the same approach. The variables we set equal to 0 and those we solve for are given special names:

- **Nonbasic variables** are those set equal to 0.
- **Basic variables** are those not necessarily equal to 0 (the ones we solve for).

Likewise, in the graphical interpretation of the simplex method, we decided to increase x_2 first because it had the largest coefficient in the objective function. This variable and its corresponding column in the tableau are given special names:

- The **pivot column** is the column with the largest negative coefficient in the bottom row of the tableau.
- The **entering basic variable** is the variable corresponding to the pivot column.

In this case the pivot column is column 2 (corresponding to x_2). Initially the nonbasic variables are the decision variables. Every variable will be either basic or nonbasic, so the initial basic variables are the slack variables and z. Each row of the tableau has an associated basic variable (the basic variable with a coefficient of 1).

Step 2: Identify the basic variable for each row of the tableau by entering it in the left-hand column of the tableau.

In the graphical interpretation of the simplex method, we decided how much we could increase x_2 by dividing the RHS of each constraint by the corresponding coefficient of x_2. We perform a similar step here.

Step 3: Compute the ratio of the RHS to the coefficient in the pivot column for the top two rows of the tableau. This can be done by entering the formulas in Figure 7.20.

Ratio
=I5/E5
=I6/E6

Figure 7.20

In the graphical interpretation of the simplex method, the smallest ratio that we just computed told us how much we could increase x_2. The row of the corresponding constraint is given a special name:

- The **pivot row** is the row with the smallest positive ratio of RHS to coefficient in the pivot column.
- The **leaving basic variable** is the variable corresponding to the pivot row.
- The **pivot** is the entry at the intersection of the pivot column and the pivot row.

   Step 4: Identify the pivot column, pivot row, and the leaving basic variable by entering them to the right of the tableau.

At this point, the initial tableau should look like Figure 7.21.

Tableau 0

Basic	x_1	x_2	y_1	y_2	z	RHS	Ratio
y1	3	2	1	0	0	200	100
y2	1	2	0	1	0	100	50
z	-3	-4	0	0	1	0	

Pivot Column =	2
Pivot Row =	2
Leaving Basic Variable =	y2

Figure 7.21

Step 5: Do elementary row operations to the matrix of coefficients so the pivot is 1 and all other entries in the pivot column are 0 (called "clearing the pivot column").

Step 5 can be done by entering the formulas from Figure 7.22 into Tableau 1.

	D	E	F	G	H	I
11	x_1	x_2	y_1	y_2	z	RHS
13	=D5-2*D14	=E5-2*E14	=F5-2*F14	=G5-2*G14	=H5-2*H14	=I5-2*I14
14	=D6/2	=E6/2	=F6/2	=G6/2	=H6/2	=I6/2
15	=D7+4*D14	=E7+4*E14	=F7+4*F14	=G7+4*G14	=H7+4*H14	=I7+4*I14

Figure 7.22

Step 6: Repeat Steps 2–5 until there are no more negative entries in the bottom row.

After repeating Steps 2–4, Tableau 1 should look like Figure 7.23.

Tableau 1

Basic	x_1	x_2	y_1	y_2	z	RHS	Ratio
y1	2	0	1	-1	0	100	50
x2	0.5	1	0	0.5	0	50	100
z	-1	0	0	2	1	200	

Pivot Column =	1
Pivot Row =	1
Leaving Basic Variable =	y1

Figure 7.23

Applying Step 5 to Tableau 1 results in Tableau 2 shown in Figure 7.24.

Tableau 2

Basic	x_1	x_2	y_1	y_2	z	RHS	Ratio
x2	1	0	0.5	-0.5	0	50	
x1	0	1	-0.25	0.75	0	25	
z	0	0	0.5	1.5	1	250	

Figure 7.24

In Tableau 2 we see that there is no negative coefficient in the bottom row, so there is no pivot column. Therefore we are done. The nonbasic variables (the ones set equal to 0) are y_1 and y_2, so the tableau gives the optimal solution shown in Figure 7.25. This is the exact same solution found with other methods. The numbers 0.5 and 1.5 in the bottom row have special meanings, as we will see in the next section.

x_1	x_2	y_1	y_2	z
50	25	0	0	250

Figure 7.25

Exercises

Directions: Solve each linear program below using the simplex method in the worksheet **2 Variable Simplex Method** or **3 Variable Simplex Method**. Verify your solution with Solver.

7.6.1

$$\text{Maximize } z = 14x_1 + 16x_2$$
$$\begin{aligned}\text{Subject to} \quad x_1 + \quad x_2 &\leq 100 \\ 20x_1 + 30x_2 &\leq 2400 \\ x_1, x_2 &\geq 0\end{aligned}$$

7.6.2

$$\text{Maximize } z = 4x_1 + 3x_2 + 6x_3$$
$$\begin{aligned}\text{Subject to} \quad 3x_1 + \quad x_2 + 3x_3 &\leq 30 \\ 2x_1 + 2x_2 + 3x_3 &\leq 40 \\ x_1, x_2, x_3 &\geq 0\end{aligned}$$

7.6.3

$$\text{Maximize } z = \quad x_1 + 2x_2 + 4x_3$$
$$\begin{aligned}\text{Subject to} \quad 3x_1 + \quad x_2 + \quad 5x_3 &\leq 10 \\ x_1 + 4x_2 + \quad x_3 &\leq 8 \\ 2x_1 \quad\quad + \quad 2x_3 &\leq 7 \\ x_1, x_2, x_3 &\geq 0\end{aligned}$$

7.7 Sensitivity Analysis

Again, consider the fruit basket problem modeled by

$$\text{Maximize } P = 3x + 4y$$
$$\begin{aligned}\text{Subject to} \quad 3x + 2y &\leq 200 \\ x + 2y &\leq 100 \\ x, x &\geq 0\end{aligned}$$

where $x =$ the number of small baskets and $y =$ the number of large baskets to produce. This has an optimal solution of $x = 50$ and $y = 25$ with a maximum profit of $250.

In doing sensitivity analysis we analyze two questions:

1. How much can a unit profit change (i.e., a coefficient in the objective function) and the optimal solution still remain optimal?

2. How much will increasing the amount of a resource increase the maximum profit?

These two questions are important economically. If the profit for an item suddenly changes, management will want to adjust the production schedule to ensure they are still maximizing profit. They need to know when this is appropriate. Also, management may want to invest money in increasing some resource (e.g., hire more employees, modernize machinery). They need to know whether this will indeed increase profit, and if so, by how much. Another reason these questions are important is that the coefficients in the problem are often *estimates*. Answering these questions can help us understand how important the accuracy of the estimates are.

Changing Unit Profits

We begin by graphically analyzing what happens to the optimal solution when a coefficient in the objective function changes. Enter the program in the worksheet **Graphical Solution**. Change the unit profit for a small basket (the coefficient of x in the objective function) to values above and below 3 and keep the profit for a large basket fixed at 4. For each one, find the optimal solution.

Figure 7.26 shows three such examples (where the dashed lines are the constraint lines and the solid lines are the level curves corresponding to the three different objective functions). Note that when the unit profit is 2.75 or 4, the optimal solution remains at $(50, 25)$ (the maximum value of P is different in each case, though). When the unit profit is 9, the optimal solution changes to $(66.67, 0)$. In general we see that for small changes in the unit profit, the solution remains the same. If the change is large enough, the solution changes. We make the same observations if we change the unit profit for a large basket while keeping the profit for a small basket fixed at 3.

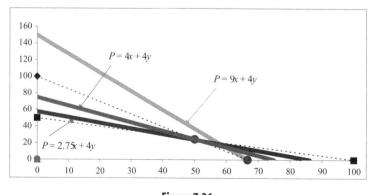

Figure 7.26

Graphically we observe that as long as the slope of the profit line is between the slopes of the constraint lines, the optimal solution will remain the same. To quantify this, we calculate the slopes of the constraint lines:

Constraint 1: $3x + 2y = 200 \Rightarrow y = -\frac{3}{2}x + 100$

Constraint 2: $x + 2y = 100 \Rightarrow y = -\frac{1}{2}x + 50$

Therefore, as long as the slope of the profit line is between $-\frac{3}{2}$ and $-\frac{1}{2}$, the optimal solution will remain the same.

Now suppose the unit profit for small baskets is C. Then the profit equation is

$$P = Cx + 4y \Rightarrow y = -\frac{C}{4}x + \frac{P}{4}.$$

Thus we see that the slope of the profit line is determined by only the value of C. The value of the profit P does not affect the slope, only the y-intercept. Therefore, if

$$-\frac{3}{2} \leq -\frac{C}{4} \leq -\frac{1}{2} \Rightarrow 2 \leq C \leq 6,$$

then the optimal solution will remain at $(50, 25)$. However, note that this is valid only if the unit profit for large baskets remains fixed at 4.

Now let K represent the unit profit for large baskets. The profit equation is

$$P = 3x + Ky \Rightarrow y = -\frac{3}{K}x + \frac{P}{K}.$$

Again we see that the slope of the profit equation is determined by only the value of K. Therefore, if

$$-\frac{3}{2} \leq -\frac{3}{K} \leq -\frac{1}{2} \Rightarrow 2 \leq K \leq 6,$$

then the optimal solution will remain at $(50, 25)$. Similar to before, this is valid only if the unit profit for small baskets remains fixed at 3.

Thus the answer to the first question is this: The unit profit for a small basket can decrease as much as $1 or increase as much as $3. The unit profit for a large basket can change as much as $2. In either case the solution $(50, 25)$ will remain optimal. This is valid only if one unit profit changes. If they both change, then we must solve the problem again (or do more analysis).

Tableau 2

Basic	x_1	x_2	y_1	y_2	z	RHS	Ratio
x2	1	0	0.5	-0.5	0	50	
x1	0	1	-0.25	0.75	0	25	
z	0	0	0.5	1.5	1	250	

Figure 7.27

Increasing Resources

We have already found the answer to the second question. It is given to us in the last row of the final tableau from the simplex method shown in Figure 7.27.

The numbers 0.5 and 1.5, the coefficients of the slack variables in the bottom row of the final tableau, are called the *shadow prices* of the apple and orange resources, respectively. These numbers tell us that increasing the number of apples by one unit will increase the maximum profit by $0.50 and increasing the number of oranges by one unit will increase the maximum profit by $1.50.

To understand why these shadow prices have these meanings, observe that the bottom row of tableau 2 is related to tableau 0 by the formula

$$(\text{Bottom row of tableau 2}) = 0.5 \times (\text{row 1 of tableau 0})$$
$$+ 1.5 \times (\text{row 2 of tableau 0})$$
$$+ (\text{row 3 of tableau 0}).$$

Therefore, if the RHS of row 1 of tableau 0 is increased by 1 (i.e., the number of apples is increased by 1), then the RHS of the bottom row of tableau 2 is increased by 0.5 (i.e., the maximum profit is increased by 0.5). Likewise, if the RHS of row 2 of tableau 0 is increased by 1, then the RHS of the bottom row of tableau 2 is increased by 1.5.

In general this means that if the number of apples is changed by n units (positive or negative), then

$$\text{Change in maximum profit} = 0.5n. \tag{7.4}$$

Note that this conclusion holds *only* if we change *only* the number of apples.

However, this conclusion is valid only if the number of apples changes within limits. To illustrate this, graph the feasible region of the program (with the original unit profits) in the worksheet **Graphical Solutions**. Change the number of apples (the RHS of constraint 1) to values above and below 200. Each time, find the optimal solution. Three such examples are given in Figure 7.28 where the solid line is the constraint 2 line, the dashed lines are the constraint 1 lines for three different numbers of apples, and the solid dots are the optimal solutions for the three cases.

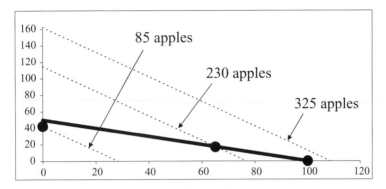

Figure 7.28

Notice that for 230 apples, the optimal solution still lies at the intersection of the constraint 1 and constraint 2 lines (it is different than the original solution, but it still occurs at this intersection). For 325 apples, the number of apples is not a constraint and the optimal solution occurs at the x-intercept of the constraint 2 line. For 85 apples, the number of oranges is not a constraint and the optimal solution occurs at the y-intercept of the constraint 1 line. Thus we see that if the number of apples changes enough, the nature of our solution changes. This means that equation (7.4) holds over only a limited domain.

Note graphically that for 325 apples the x- and y-intercepts of the constraint 1 line are both greater than the corresponding intercepts of the constraint 2 line. For 85 apples, the opposite is true. Based on this observation we can conclude that the optimal solution will occur at the intersection of the constraint 1 and constraint 2 lines only if this relation among the intercepts does not hold.

To calculate the domain over which equation (7.4) holds, note that the constraint 2 line is given by $x + 2y = 100$, which has x- and y-intercepts of 100 and 50, respectively. Now suppose we have A apples available. Then the constraint 1 line is given by $3x + 2y = A$, which has x- and y-intercepts of $\frac{A}{3}$ and $\frac{A}{2}$, respectively.

The intercepts of the constraint lines are equal when

$$x\text{-intercepts: } 100 = \tfrac{A}{3} \quad \Rightarrow \quad A = 300$$
$$y\text{-intercepts: } \ \ 50 = \tfrac{A}{2} \quad \Rightarrow \quad A = 100.$$

Thus we conclude that equation (7.4) holds when the number of apples is between 100 and 300 (or n is between -100 and $+100$). The economic interpretation of this is

If each apple costs *less* than $0.50, the manager can increase the number of apples by up to 100 to increase the maximum profit. If each apple can be sold for *more* than $0.50, up to 100 apples should be sold. This will decrease the profit from fruit baskets, but increase the total profit for the produce department.

Now let's fix the number of apples at 200 and change the number of oranges. The constraint 1 line is given by $3x + 2y = 200$, which has x- and y-intercepts of $\frac{200}{3}$ and 100, respectively. Suppose we have B oranges available. Then the constraint 2 line is given by $x + 2y = B$, which has x- and y-intercepts of B and $\frac{B}{2}$, respectively.

The intercepts of the constraint lines are equal when

$$x\text{-intercepts: } \frac{200}{3} = B \quad \Rightarrow \quad B \approx 66.7$$
$$y\text{-intercepts: } 100 = \frac{B}{2} \quad \Rightarrow \quad B = 200.$$

This means that up to 100 oranges may be added and the maximum profit will increase by $1.50 (the shadow price for oranges) for each one. Up to 33 oranges may be removed and the maximum profit will decrease by $1.50 for each one.

Solver will generate a sensitivity analysis report after it solves a model. To examine this report for this program, return to the worksheet **Fruit Baskets** created in Section 7.2 (see Figure 7.1). Set the appropriate Solver parameters (be sure to select **Assume Linear Model** under **Options**) and click **Solve**. In the next window that appears, select **Sensitivity** under **Reports** and click **OK**. You should get the new worksheet containing the results shown in Figure 7.29.

A	B	C	D	E	F	G	H
6	Adjustable Cells						
7			Final	Reduced	Objective	Allowable	Allowable
8	Cell	Name	Value	Cost	Coefficient	Increase	Decrease
9	$B$2	Number Small	50	0	3	3	1
10	$C$2	Number Large	25	0	4	2	2
11							
12	Constraints						
13			Final	Shadow	Constraint	Allowable	Allowable
14	Cell	Name	Value	Price	R.H. Side	Increase	Decrease
15	$D$3	Apples Amt Used	200	0.5	200	100	100
16	$D$4	Oranges Amt Used	100	1.5	100	100	33.33333333

Figure 7.29

Cell **F9** shows the unit profit for small baskets. Cells **G9** and **H9** tell us that this unit profit can increase by as much as 3 or decrease as much as 1 (i.e., varies between 2 and 6) and the optimal solution will remain the same. Row 10 gives similar results for large baskets.

Cell **E15** gives the shadow price for apples. Cell **F15** shows the number of available apples. Cells **G15** and **H15** tell us that the interpretation of the shadow price is valid if the number of apples is increased by 100 or decreased by 100 (i.e., is between 100 and 300). Row 16 gives similar results for the oranges.

These are the exact same conclusions we reached earlier.

Exercises

7.7.1 Suppose the manager can buy a box of 50 oranges or a box of 50 apples for making fruit baskets for $20 each, but he can purchase only one. Which box should he purchase? Why?

7.7.2 Suppose there are 350 apples and 100 oranges available for making fruit baskets. Generate a sensitivity report and explain why the shadow price for apples is 0, the allowable increase is ∞, and the allowable decrease is only 50.

7.7.3 A toy company manufactures cars and trucks. Each car uses 1 unit of plastic and 20 units of metal and yields a profit of $14. Each truck uses 1 unit of plastic and 30 units of metal and yields a profit of $16. The company has 100 units of plastic and 2400 units of metal available. To determine the number of each toy to manufacture to maximize profit, we need to solve the program

$$\textbf{Maximize } P = 14x_1 + 16x_2$$
$$\textbf{Subject to} \qquad x_1 + \quad x_2 \le 100$$
$$20x_1 + 30x_2 \le 2400$$
$$x_1, x_2 \ge 0$$

where x_1 = the number of cars and x_2 = the number of trucks to produce. This is the same program we solved in Exercise 7.6.1.

1. Use the final tableau in your solution to Exercise 7.6.1 to find the shadow prices of the cars and the trucks.

2. By hand, find the range over which the interpretation of the shadow price is valid for each resource.

3. By hand, find ranges for the unit profits over which the solution remains optimal.

4. Verify your calculations with Solver.

7.8   The Gradient Method

A *nonlinear program* is any program that does not fit the definition of a linear program. A linear program has a very special structure that can be utilized to construct an efficient solution method (e.g., the simplex method). The drawback of this is that if a problem does not fit this structure, the solution method cannot be applied.

Since nonlinear programs do not fit the structure of a linear program, the simplex method cannot solve them. In fact, there is no efficient algorithm that guarantees an optimal solution to a general nonlinear program. However, certain types of nonlinear programs— such as binary integer, convex, separable, and quadratic—have special structures that can be utilized to find efficient solution methods.

In this section we examine a numerical approach used by Solver to approximate the solution to a nonlinear program called the *gradient method*.

Example 7.8.1   Calculus I Problem

Consider the following unconstrained nonlinear program with one decision variable:

$$\textbf{Maximize } f(x) = -x^2 + 4x.$$

This is a typical maximization problem from Calculus I. The first step is to set the first derivative equal to 0 and solve to find the critical points:

$$f'(x) = -2x + 4 \overset{\text{SET}}{=} 0 \quad \Rightarrow \quad x = 2.$$

Critical points are possible values of x at which the function has a *local* (or *relative*) *maximum* or *minimum*. A local maximum occurs at x_0 if

$$f(x_0) \geq f(x) \text{ for all } x \text{ in some interval centered at } x_0.$$

A local minimum is defined similarly. By contrast, a *global maximum* occurs at x_0 if

$$f(x_0) \geq f(x) \text{ for all } x \text{ in the domain of } f.$$

A global minimum is also defined similarly. As an example of these definitions, see the graph of $y = g(x)$ shown in Figure 7.30. The function g has a local maximum near 0.6, a global maximum near 3.5, a local minimum near 2, and no global minimum.

In a maximization problem such as this we are looking for a value of x that gives a global maximum. To determine if the critical point $x = 2$ gives a local minimum or maximum, we look at the second derivative of f at $x = 2$:

$$f''(x) = -2 \quad \Rightarrow \quad f''(2) = -2.$$

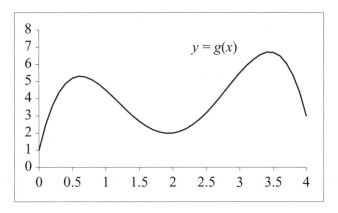

Figure 7.30

Since this second derivative is negative we conclude that f is concave downward. So by the second derivative test, f has a local maximum at $x = 2$. Since f has only one critical point, we conclude that f has a global maximum at $x = 2$.

Our solution is $x = 2$, which gives an objective function value of $f(2) = 4$.

Example 7.8.2

Consider the following unconstrained nonlinear program with one decision variable:

$$\text{Minimize } f(x) = x^4 - 8x^3 + 20x^2 - 16.5x + 7.$$

The graph of this function f is given in Figure 7.31. We see that f has a local minimum near 0.6 and a global minimum near 3.4. In this problem, we are asking for the location of the global minimum.

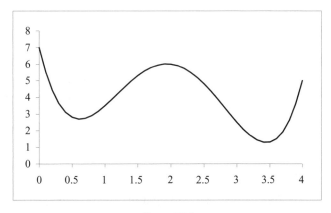

Figure 7.31

We could use the technique in Example 7.8.1 to solve this problem, but finding the critical points would involve finding the roots of a third-degree polynomial—not an easy task. To find a better approximate solution we will try to use Solver. Rename a blank worksheet "**Solver**," and format it as in Figure 7.32. Set the Solver parameters to minimize cell **B2** by

	A	B
1	x	f(x)
2	0	=A2^4-8*A2^3+20*A2^2-16.5*A2+7

Figure 7.32

changing cell **A2**. Do not add any constraints and *Do not* tell it to assume a linear model. The solution it gives is shown in Figure 7.33. Notice that it found the location of the local minimum. Next set the value of x to 4 (i.e., enter 4 in cell **A2**) and resolve it. The results

	A	B
1	x	f(x)
2	0.618138	2.699114

Figure 7.33

are shown in Figure 7.34. This time Solver found the location of the global minimum. To

	A	B
1	x	f(x)
2	3.444485	1.285246

Figure 7.34

understand why Solver does not always find the location of the global minimum, we need to examine the basic idea behind the gradient method.

One-Dimensional Gradient Method

First we will examine the gradient method applied to minimizing a function $f(x)$ with one variable. The gradient method is an iterative approach for finding a critical point of the function (i.e., a value of x for which $f'(x) = 0$).

A simplified version of the gradient method to minimize a function $f(x)$ is given in the following algorithm:

1. Choose an initial value, x_0.
2. Let $x_{k+1} = x_k - \lambda f'(x_k)$ where $\lambda > 0$ is some specified constant.
3. Repeat Step 2 for 50 iterations (50 iterations is arbitrary).

The basic idea behind the gradient method is that if the derivative $f'(x_k)$ is positive, then the function is increasing, so we want to decrease x_k. If $f'(x_k)$ is negative, then the function is decreasing, so we want to increase x_k.

The constant λ affects how much x_k is changed in each step. In more sophisticated versions, the algorithm would terminate when $|f'(x_k)|$ is sufficiently close to 0, rather than arbitrarily after 50 iterations. Ideally, when the algorithm terminates, x_k is near a critical point. We could modify this algorithm to maximize a function by changing the minus sign in Step 2 to a plus sign.

To implement this simple algorithm in Excel and to minimize the function $f(x) = x^4 - 8x^3 + 20x^2 - 16.5x + 7$ introduced in Example 7.8.1, follow these steps:

1. First we will create a graph of $f(x)$ over the interval $0 \le x \le 4$. Rename a blank workbook "**1-Dim**" and format it as in Figure 7.35. Copy row 3 down to row 42.

	A	B
1	x	f(x)
2	0	=A2^4-8*A2^3+20*A2^2-16.5*A2+7
3	=A2+0.1	=A3^4-8*A3^3+20*A3^2-16.5*A3+7

Figure 7.35

2. Create a graph similar to Figure 7.36.

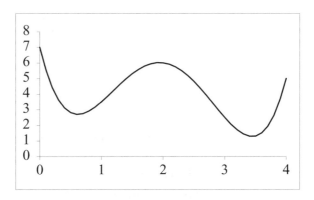

Figure 7.36

3. Add the formulas in Figure 7.37 and copy row 5 down to row 53.

	D	E	F	G
1	$\lambda =$	0.115		
2				
3	k	x_k	$f(x_k)$	$f'(x_k)$
4	1	0	=E4^4-8*E4^3+20*E4^2-16.5*E4+7	=4*E4^3-24*E4^2+40*E4-16.5
5	=D4+1	=E4-G4*$E$1	=E5^4-8*E5^3+20*E5^2-16.5*E5+7	=4*E5^3-24*E5^2+40*E5-16.5

Figure 7.37

4. To visualize how this algorithm works, add a scroll bar, and link it to cell **J2** with a **min** and **max** of **1** and **50**, respectively. Add the formulas in Figure 7.38.

	I	J	K	L
1			x	f(x)
2	k=	1	=OFFSET(E3,J2,0)	=OFFSET(F3,J2,0)

Figure 7.38

5. Drag the cells **K2:L2** onto the graph, and add the cells as a new series with x-values in the first column. Right-click on the point on the graph you just added, and format it so the graph resembles Figure 7.39.

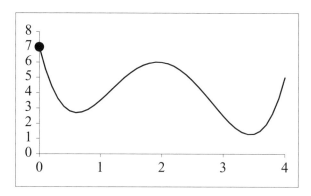

Figure 7.39

Move the slider back and forth and notice how the point on the graph eventually settles to one of the two local minima on the graph. Change the initial value of x and the value of λ and note how they affect the search and the solution. Note that if λ is large (e.g., $\lambda = 0.2$), the point "jumps" around and never settles at a local minimum. If λ is too small, the point doesn't move toward a local minimum very quickly. Also note that sometimes we get the local minimum at $x = 0.62$ and other times we get the global minimum at $x = 3.45$.

Two-Dimensional Gradient Method

Suppose $f(x, y)$ is a differentiable function of two variables, x and y. The gradient vector ∇f is defined as

$$\nabla f = \begin{bmatrix} \frac{\partial f}{\partial x} \\ \frac{\partial f}{\partial y} \end{bmatrix}.$$

The two-dimensional gradient method is based on the fact that the gradient vector ∇f at a point in the domain of f always points in the direction of the maximum rate of *increase* of the function, and that at a point of *local* minimum or maximum, (x_c, y_c) (called a *critical point*),

$$\nabla f(x_c, y_c) = \begin{bmatrix} 0 \\ 0 \end{bmatrix}.$$

As in the one-dimensional case, the gradient method can, at best, approximate a critical point of the function $f(x, y)$. A simple gradient method algorithm for minimizing a function $f(x, y)$ is given below:

1. Choose an initial value (x_0, y_0).

2. Let $\begin{cases} x_{k+1} = x_k - \lambda \frac{\delta f}{\delta x}(x_k, y_k) \\ y_{k+1} = y_k - \lambda \frac{\delta f}{\delta y}(x_k, y_k) \end{cases}$ where $\lambda > 0$ is some specified constant.

3. Repeat Step 2 for 50 iterations (50 is arbitrary).

The basic idea behind this algorithm is the same as for the one-dimensional case. This algorithm could be changed to maximize a function by changing the minus signs in Step 2 to plus signs. In more sophisticated versions, the algorithm terminates when the norm of the gradient, $\|\nabla f(x_k, y_k)\|$, is sufficiently close to 0.

To illustrate this algorithm, consider the unconstrained two-dimensional program

Minimize $f(x, y) = x^4 + y^4 - 4xy + 2.$

For this function,

$$\frac{\delta f}{\delta x} = 4x^3 - 4y$$

$$\frac{\delta f}{\delta y} = 4y^3 - 4x.$$

Follow these steps to implement the gradient method for this function:

1. Rename a blank worksheet "**2-Dim**" and format it as in Figure 7.40. Copy row 5 down to row 54.

	A	B	C	D	E	F
1		$\lambda =$ 0.02				
2						
3	k	x_k	y_k	$f(x_k, y_k)$	df/dx	df/dy
4	0	0	0.5	=B4^4+C4^4-4*B4*C4+2	=4*B4^3-4*C4	=4*C4^3-4*B4
5	=A4+1	=B4-$C$1*E4	=C4-$C$1*F4	=B5^4+C5^4-4*B5*C5+2	=4*B5^3-4*C5	=4*C5^3-4*B5

Figure 7.40

2. To get a graphical perspective of how (x_k, y_k) changes, add the formulas in Figure 7.41 to the worksheet.

	H	I	J
1	k	x_k	y_k
2	0	=OFFSET(B4,H2,0)	=OFFSET(C4,H2,0)

Figure 7.41

3. Create a graph as in Figure 7.42 using the data in the range **I2:J2**. Set the x- and y-axes **min** and **max** to -2 and $+2$, respectively.

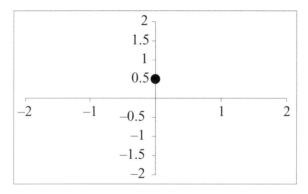

Figure 7.42

4. Add a scroll bar, set the **min** and **max** to **0** and **50**, respectively, and set the linked cell to **H2**. Move the slider back and forth and watch the graph in Figure 7.42. Note that as the algorithm proceeds, (x_k, y_k) gets closer to $(1, 1)$.

5. Use the data in columns titled **k** and **f(x, y)** to create a graph as shown in Figure 7.43. This graph shows that as the algorithm proceeds, the value of $f(x_k, y_k)$ decreases from 2 to 0. This is exactly what we want. These results indicate that there is a local minimum of 0 at the point $f(1,1)$.

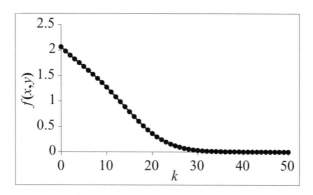

Figure 7.43

6. Try the additional initial values listed in Table 7.21.

Initial Value	Result
$(1,1)$	The gradient is exactly 0, so (x_k, y_k) doesn't change.
$(0.1, 0)$	Again, (x_k, y_k) moves toward $(1,1)$.
$(0,0)$	The gradient is exactly 0, but $f(0,0) = 2$, so $(0,0)$ is not a location of the global minimum.
$(-2, -2)$	Note that (x_k, y_k) moves toward $(-1,-1)$ and that $f(-1,-1) = 0$.
$(-1,1)$	Here, (x_k, y_k) moves toward $(0,0)$. This indicates that $(0,0)$ is a possible location of a local minimum. This is often called "getting stuck at a local minimum."

Table 7.21

Summarizing our results from Table 7.21, it appears that this function has a global minimum of 0 at $(1,1)$ and at $(-1,1)$, and that there is a critical point at $(0,0)$. Especially note that the result is affected by the initial value. The algorithm doesn't always move towards the *global* minimum.

Now let's examine how Solver handles this same minimization problem. Rename a blank worksheet "**Solver 2**" and format it as in Figure 7.44.

	A	B	C
1	Solution		
2	**x**	**y**	**f(x,y)**
3	0	0.5	=A3^4+B3^4-4*A3*B3+2

Figure 7.44

Format Solver to minimize cell **C3** by changing cells **A3:B3**. Do not add any constraints and *do not* tell it to assume a linear model. Notice that Solver found the solution $(1, 1)$, which is indeed the location of the global minimum. However, if we enter the different initial values in Table 7.21 in the range **A3:B3** and rerun Solver, note that Solver does not always find the global minimum. It can, and does, get stuck at a local minimum. This illustrates that we must be very careful about choosing the initial values. We might want to try several initial values.

The above discussion illustrates that the gradient method is a rather simple algorithm for approximating critical points of a function. The drawback is that it cannot determine whether the solution it finds is a local or global minimum or maximum without some additional external knowledge of the function. Therefore, it, along with Solver, must be used with caution when solving nonlinear programs.

Exercises

7.8.1 Use 100 iterations of the gradient method to approximate the local *minimum* value of the function $f(x, y) = 2x^2 + 6xy + 6y^2 - 3x + 5y$. What happens if λ is too large? What if it is too small?

7.8.2 Use 100 iterations of the gradient method to approximate the local *maximum* value of the function $f(x, y) = 90x - 0.1x^2 + 15y - 0.15y^2 - 0.05xy - 2000$. What happens if λ is too small?

7.8.3 The distance the gradient method moves the point (x_k, y_k) to (x_{k+1}, y_{k+1}) is affected by two quantities: the length of the gradient, $\|\nabla f(x_k, y_k)\|$, and the value of λ. As the algorithm proceeds, $\|\nabla f(x_k, y_k)\|$ gets smaller and smaller, so the point does not move as far. As we have seen, if λ is too small, the algorithm doesn't find the optimal solution in a reasonable number of iterations.

One way to get around this problem is to increase the value of λ in each iteration. A simple algorithm for maximizing a function that incorporates this idea is given below:

1. Choose initial values (x_0, y_0) and λ_0.

2. Let
$$\begin{cases} x_{k+1} = x_k + \lambda_k \frac{\delta f}{\delta x}(x_k, y_k) \\ y_{k+1} = y_k + \lambda_k \frac{\delta f}{\delta y}(x_k, y_k). \end{cases}$$

3. Let $\lambda_{k+1} = \delta\lambda_k$ where $\delta > 1$ is some specified constant.

4. Repeat Steps 2 and 3 for 100 iterations (100 is arbitrary).

Implement this algorithm to maximize the function in Exercise 7.8.2 using $\lambda_0 = 0.1$ and $\delta = 1.1$. How many iterations are really necessary for the gradient to get reasonably close to $(0,0)$? What happens if we let the algorithm go for too many iterations?

For Further Reading

- For a classic reference on everything related to operations research, see F. Hillier and G. Lieberman, 2001, *Introduction to Operations Research*, 7th ed., McGraw-Hill, Boston, MA.

- For more information on the simplex method, sensitivity analysis, nonlinear programming, and other topics from this chapter, see W. L. Winston and M. Venkataramann, *Introduction to Mathematical Programming, Operations Research: Volume One*, 4th ed., Thomson Brooks/Cole, Pacific Groves, CA.

Reference

Hillier, F., and G. Lieberman. 2001. *Introduction to operations research*. 7th ed. Boston, MA: McGraw-Hill, 382.

APPENDIX A

Spreadsheet Basics

Here we explain some of the basic terminology and tools used to build the models in this book. These explanations apply directly to Microsoft Office Excel 2003, although most of them apply to other versions of Excel as well. There is a free spreadsheet program similar to Excel called OpenOffice.org Calc that is available for download as part of the OpenOffice.org suite at *http://www.openoffice.org/*. Many of the commands used in OpenOffice.org Calc are similar to those in Excel, but some are significantly different. Almost all the models in this book can be built with OpenOffice.org Calc.

A.1   Basic Terminology

When you first open Excel, you should get a window that looks similar to Figure A.1. (Your window will probably not look exactly like that in Figure A.1 due to the placement of toolbars and icons.)

Each rectangle, called a *cell*, is a place where data, text, or formulas can be entered. A collection of cells is called a *worksheet*. The name of the worksheet is given in the *worksheet tab* near the bottom of the worksheet. The worksheet name can be changed by right-clicking on the worksheet tab and selecting **Rename**. A collection of worksheets is called a *workbook*. Worksheets can be added or deleted from a workbook by right-clicking on the worksheet tab and selecting either **Insert...** or **Delete**. A worksheet tab can be moved to the left or right by left-clicking, holding, and dragging.

The name of each *column* is listed along the top of the worksheet while the number of each *row* is listed along the left-hand side of the worksheet. The width or height of a column or row can be changed by left-clicking and holding on the right or bottom and then dragging

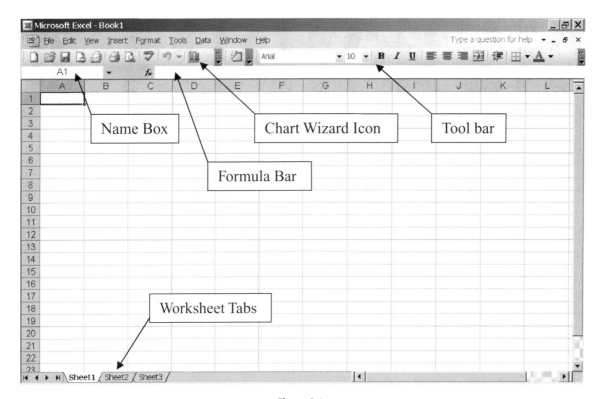

Figure A.1

it to the desired width or height. A cell is named according to its column and row position. The *selected cell* has a thicker border around it and its name is shown in the *name box*. The selected cell can be changed using the arrow keys or by clicking on another cell.

A two-dimensional *range* of cells can be selected by left-clicking on the cell in the upper-left corner of the range, holding, and dragging to the cell lower-right corner of the range. This highlights these cells, indicating they are all selected. Ranges are referred to by the cells in the upper-left and lower-right corners in the form **(Upper Left):(Lower Right)**.

The *formula bar* displays the contents of the selected cell. The *chart wizard icon* is contained in one of the *toolbars* and is a link to the chart wizard, which is the tool for creating charts and graphs. Most of the models in this book require a chart of one type or another, so this icon is handy to have readily available in the toolbar. (If this icon is not available, select **View** → **Toolbars** → **Customize...** → **Commands** → **Insert** and drag the chart wizard icon onto the toolbar.) More details on using the chart wizard will be given later.

A.2   Entering Text, Data, and Formulas

Text, data, and formulas are easily entered by selecting the desired cell, typing the desired contents, and pressing **Enter**. To practice doing this, format a blank worksheet as in Figure A.2. This worksheet contains two columns of data named "**x**" and "**y**" and a third column named "**z**" that we will define later.

	A	B	C
1	x	y	z
2	1	5	
3	2	8	

Figure A.2

Notice that when you press Enter, the selected cell moves to the cell directly under the previous one. By default text is left-justified. The text in row 1 can be changed to bold and centered by selecting the range **A1:A3**, and then clicking on the bold font icon and then the center icon located in the toolbar.

Now suppose we want to define the quantity **z** to be **x** + **y**. We can easily do this by entering the formula in Figure A.3. Every formula begins with an equals sign. This formula can be entered by typing it as in the figure and then pressing **Enter**, or you can type =, click on cell **A2**, type +, click on cell **B2**, and then press **Enter**.

	C
2	=A2+B2

Figure A.3

Once the formula is entered, select cell **C2** and click in the formula bar. Notice how different colored boxes are put around cells **A2** and **B2** and that the **A2** and **B2** in the formula are changed to the corresponding colors. This feature simplifies the process of debugging formulas.

To calculate the second value of **z**, we could type the formula **=A3+B3** in cell **C3**, but there is an easier way. Select cell **C2**, left-click and hold on the dark square in the lower-right corner of the cell. Then drag the box down one row and release. The results are shown in Figure A.4. This is exactly what we want.

A.2.1   Understanding Cell References

To understand why the formula in cell **C2** is copied down to cell **C3** in this way, we need to understand what we mean when we reference cells in formulas. The formula in cell **C2** should not be interpreted as "add cell **A2** to cell **B2**." Rather, it should be interpreted as "add the cell two columns to the left and in the same row to the cell one column to the left and in the same row." In other words, these cell references are *relative*. When this formula is copied down one row, the cell "two columns to the left and in the same row" is now **A3** and the cell "one column to the left and in the same row" is now **B3**.

	C
1	z
2	=A2+B2
3	=A3+B3

Figure A.4

Now, change the formula in cell **C2** to that shown in Figure A.5. The **$** symbols can be entered manually or you can delete the contents of **C2**, then type =, click on cell **A2**, press the **F4** key, type +, click on cell **B2**, press the **F4** key, and then press **Enter**.

	C
2	=$A$2+$B$2

Figure A.5

Copy the formula in cell **C2** down to **C3**. The results are shown in Figure A.6. Notice that the formula did not change. This is because the **$** symbols "fix" the row and column reference. So the formula in **C2** really does mean "add cell **A2** to cell **B2**." When we copy it down, the meaning does not change.

	C
1	z
2	=$A$2+$B$2
3	=$A$2+$B$2

Figure A.6

Now, change the formula in cell **C2** to that shown in Figure A.7. The **$** symbols can be manually entered or they can be entered by selecting the cells and clicking **F4** two or three times, similar to the process used earlier.

	C
1	z
2	=A$2+$B2

Figure A.7

Copy the formula in cell **C2** down one row and to the right one column. The results are shown in Figure A.8.

	C	D
1	z	
2	=A$2+$B2	=B$2+$B2
3	=A$2+$B3	

Figure A.8

We get these results because the **$** in **A$2** "fixes" the row at 2, but the column is still relative. When we copy down, this row does not change, but when we copy to the right, the column changes to **B**. Likewise, the **$** in **$B2** "fixes" the column at **B**, but the row is still relative. When we copy down, this row changes, but when we copy to the right, the column does not change.

A.2.2   Formatting Cells

The formats of a cell or range can be easily changed by first selecting the cell or range and then right-clicking within the cell or range. Selecting **Format Cells...** yields the window shown in Figure A.9.

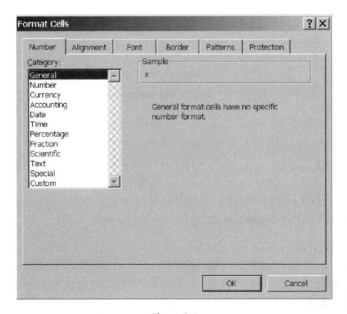

Figure A.9

Several of these tabs are useful for building the models in this book:

- **Number**—The Number tab allows you to change the way numbers are displayed. For instance, selecting **Number** under **Category:** allows you to, among other things, set the number of displayed decimal places.
- **Font**—The Font tab allows you to change the font, font style, and size of text. It also allows you to add effects such as superscript or subscript.
- **Border**—The Border tab allows you to change the border around and between cells.
- **Patterns**—The Patterns tab allows you to change the background color and pattern of cells.

A.3   Creating Charts and Graphs

To illustrate the process of creating charts and graphs, format a blank worksheet as in Figure A.10.

	A	B
1	x	y
2	1	2
3	2	5
4	3	9
5	4	12
6	5	13

Figure A.10

To create a simple plot of **y** versus **x**, follow these steps:

1. Select the range **A1:B6** and click on the chart wizard icon. Selecting **XY (scatter)** under **Chart type:** yields the window shown in Figure A.11.

   The left-hand side of the window shows the different types of charts that are possible. In this book we most frequently use the **XY (scatter)** chart. The right-hand side shows the different subtypes. The box under the subtypes gives a brief description of the selection. For this demonstration, choose the subtype on the top and click **Next**.

2. Selecting the **Series** tab in the next window yields the window shown in Figure A.12. Typically, nothing needs to be changed in this window. The ranges of the **X-values** and the **Y-values** (corresponding to the horizontal and vertical axes, respectively) are given. Combined, these ranges are referred to as the **Series**. After the chart is created, this window can be used to add new series.

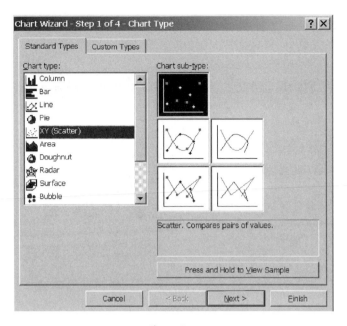

Figure A.11

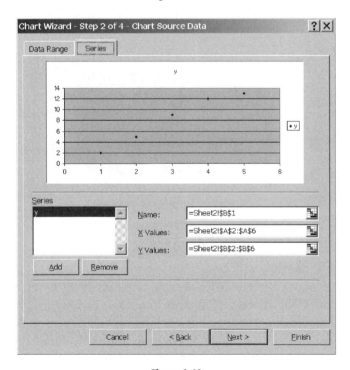

Figure A.12

3. Clicking **Next** yields a window where you can change the chart options. Set the **X-** and **Y**-axes values as shown in Figure A.13.

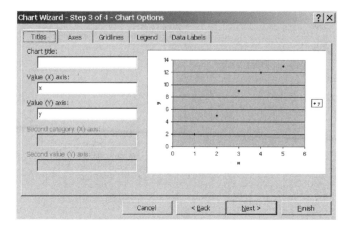

Figure A.13

4. Click **Next** and then **Finish** to yield the chart shown in Figure A.14.

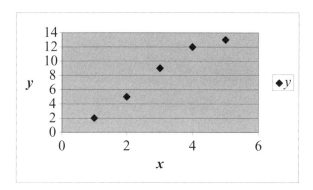

Figure A.14

5. The color of the *plot area* (the gray area) can be changed by right-clicking on it and selecting **Format Plot Area....** In this book we always use a white plot area. This can most easily be done by simply left-clicking on the plot area and pressing **Delete**. The gridlines and legend can be deleted in a similar fashion. Doing this yields the chart shown in Figure A.15.

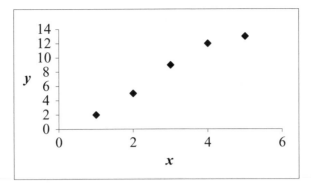

Figure A.15

6. The format of the x- and y-axes can be changed by double-clicking on an axis. The **Source Data** and **Chart Options** can be changed by right-clicking on the chart.

A.3.1   Graphing Functions

Excel does not have a built-in tool for graphing functions, but we can easily create an x–y table and then plot the points. For example, to graph the function $f(x) = x^2$ over the interval $[-2, 2]$, format a blank worksheet as in Figure A.16. Copy row 3 down to row 42.

	A	B
1	x	y
2	-2	=A2^2
3	=A2+0.1	=A3^2

Figure A.16

Select columns **A** and **B** by left-clicking and holding on the column **A** header and then dragging to column **B**. Use the chart wizard to create an XY (scatter) chart. Choose the subtype described as "Scatter with data points connected by smoothed lines without markers." Once the chart is created, delete the plot area and the gridlines; set the x-axis **min** and **max** to -2 and $+2$, respectively; and set the y-axis **min** and **max** to **0** and **4**, respectively, with a major unit of 1. The result should look like Figure A.17.

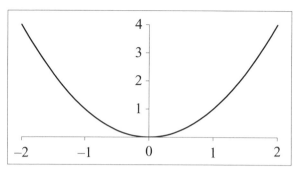

Figure A.17

A.4   Scroll Bars

Scroll bars allow us to change the value of a cell with a graphical interface. This allows us to dynamically change the values of parameters within a model and analyze the results. In a blank worksheet, select **View** → **Toolbars** → **Control Toolbox**. A window similar to that in Figure A.18 should appear.

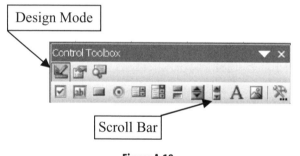

Figure A.18

For our purposes, there are only two important buttons in this window: **Design Mode** and **Scroll Bar**. When you click the scroll bar button, the cursor changes to a small cross. Use this to draw a long, skinny rectangle. Right-click on the resulting scroll bar and select **Properties**. A window similar to that in Figure A.19 should appear.

There are three important properties we need to change. The **LinkedCell** is the cell whose value we want to change. Set this to **A1** by typing **A1** in the box next to it. The **Min** and **Max** are the minimum and maximum values, respectively, of the cell. Set these to 0 and 1000, respectively. Close the properties window and click on the **Design Mode** button. The scroll bar is now ready to use. Move the slider on the scroll bar back and forth and note that the number in cell **A1** changes between 0 and 1000 in increments of 1. The scroll bar

Figure A.19

properties can be changed by clicking on the **Design Mode** button and then right-clicking on the scroll bar.

In most instances, we may want the value of a cell to change in increments other than 1. This can be accomplished using a formula that references the linked cell. For instance, enter the formula shown in Figure A.20. Move the slider back and forth and note that the number in cell **A2** changes between 0 and 100 in increments of 0.1.

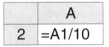

	A
2	=A1/10

Figure A.20

(**Note:** There is a somewhat easier way to create scroll bars using the **Forms** toolbar. However, these scroll bars do not work as well with graphs. With the method described previously, if the scroll bar changes a value on a graph, the graph changes in a continuous manner as the slider is moved back and forth. If the **Forms** toolbar is used, the graph will not change until you release the mouse button.)

A.5   Array Formulas

Excel can perform simple matrix operations such as addition, multiplication, and finding inverses. For example, if $A = \begin{bmatrix} 1 & 2 \\ 3 & 4 \end{bmatrix}$ and $B = \begin{bmatrix} 3 & 4 \\ 5 & 6 \end{bmatrix}$, to compute $C = A + B$, format a blank worksheet as in Figure A.21. To center the text "A" between cells **A1** and **B1**, select the range **A1:B1** and then press the **Merge and Center** icon in the toolbar. (If this icon is not available, select **Format** → **Cells...** → **Alignment**, check the box next to **Merge cells**, and click **OK**.)

	A	B	C	D	E	F	G	H
1	A			B			C	
2	1	2		5	6			
3	3	4		7	8			

Figure A.21

Next, select the range **G2:H3**, type **=A2:B3+D2:E3**, and press the combination of keys **Ctrl-Shift-Enter** (this combination tells Excel to compute an array formula). The results are shown in Figure A.22. Notice that when you select any cell in the range **G2:H3**, the formula is in curly brackets, {...}. This indicates that an array formula has been entered.

	G	H
1	C	
2	6	8
3	10	12

Figure A.22

APPENDIX B

Built-In Functions

Excel has numerous built-in functions. In this appendix, we list the functions used in this book and provide a brief description of each. Excel has a terrific help menu that describes each function in detail.

- **ABS(number)**—Returns the absolute value of **number**.
- **AND(logical1, logical2, . . .)**—Returns a 1 if all the logical statements are true and 0 otherwise.
- **AVERAGE(number1, number2, . . .)**—Returns the arithmetic mean of the list of numbers. The argument can be a range of cells.
- **COMPLEX(a, b)**—Returns an imaginary number in the form $a + bi$. If this function returns the #NAME? error, install and load the Analysis ToolPak add-in.
- **CORREL(array1, array2)**—Returns the correlation coefficient of the data in **array1** and **array2**. The two arrays are typically entered as ranges of cells.
- **COUNTIF(range, criteria)**—Returns the number of cells in the **range** that meet the given **criteria**.
- **EXP(number)**—Returns the number e raised to the power of **number**.
- **IF(logical test, value if true, value if false)**—Returns **value if true** if the **logical test** is true and **value if false** otherwise.
- **IMREAL(inumber)**—Returns the real part of the complex number **inumber**. If this function returns the #NAME? error, install and load the Analysis ToolPak add-in.
- **IMSUB(inumber1, inumber2)**—Returns the difference of the two complex numbers **inumber1** and **inumber2**. If this function returns the #NAME? error, install and load the Analysis ToolPak add-in.

- **INT(number)**—Returns **number** rounded down to the nearest integer.
- **INTERCEPT(known y's, known x's)**—Returns the y-intercept of the linear regression line fit to the data in **(known y's, known x's)**. These two lists of data are usually entered as ranges of cells.
- **LN(number)**—Returns the natural logarithm of **number**.
- **MINVERSE(array)**—Returns the inverse of the matrix in **array**. The array is usually entered as a range of cells. This is an array formula, so entering it requires the combination of keys **Ctrl-Shift-Enter**.
- **MMULT(array1, array2)**—Returns the product of the matrices in **array1** and **array2**. The two arrays are usually entered as ranges of cells. This is an array formula, so entering it requires the combination of keys **Ctrl-Shift-Enter**.
- **MOD(number, divisor)**—Returns the remainder when **number** is divided by **divisor**.
- **NA()**—Returns the error value #N/A that is used to mark empty cells.
- **NORMINV(probability, mean, standard deviation)**—Returns the value of the inverse normal cumulative distribution function with the specified **mean** and **standard deviation** and the input value of **probability**.
- **RAND()**—Returns a value of a (pseudo)random variable that is uniformly distributed over the interval $[0, 1]$.
- **RANDBETWEEN(bottom, top)**—Returns a random integer between **bottom** and **top**. If this function returns the #NAME? error, install and load the Analysis ToolPak add-in.
- **SLOPE(known y's, known x's)**—Returns the slope of the linear regression line fit to the data in **(known y's, known x's)**. These two lists of data are usually entered as ranges of cells.
- **SQRT(number)**—Returns the principal positive square root of **number**.
- **STDEV(number1, number2, . . .)**—Returns the sample standard deviation of the list of numbers. The argument can be a range of cells.
- **SUM(number1, number2, . . .)**—Returns the sum of the numbers in the list. The argument can be a range of cells.
- **SUMIF(range, criteria)**—Returns the sum of the numbers in the **range** that meet the specified **criteria**.
- **SUMPRODUCT(array1, array2, array3, . . .)**—Returns the sum of the products of corresponding entries in the arrays.

- **TRANSPOSE(array)**—Returns the transpose of the matrix in **array**. The array is usually entered as a range of cells. This is an array formula, so entering it requires the combination of keys **Ctrl-Shift-Enter**.
- **VLOOKUP(lookup value, table array, col index num)**—Searches the leftmost column of **table array** for the first value that matches **lookup value**. It returns the value in that row and column **col index num**.

Index